Mathematics in Everyday Life

John Haigh

Mathematics in Everyday Life

Second Edition

 Springer

John Haigh
Department of Mathematics
University of Sussex
Brighton, UK

ISBN 978-3-030-33086-6   ISBN 978-3-030-33087-3   (eBook)
https://doi.org/10.1007/978-3-030-33087-3

Mathematics Subject Classification (2010): 00-01, 00A05, 00A99, 34-01, 90-01, 91-01

This Springer imprint is published by the registered company Springer Nature Switzerland AG
The registered company address is: Gewerbestrasse 11, 6330 Cham, Switzerland

Preface to the Second Edition

I thank those readers who have told me of their enjoyment and illumination from the first edition. The main change now is the addition of a Chapter on computer applications, for which my colleagues James Hirschfeld and Konstantin Blyuss have been most helpful. Otherwise, the material has been updated in places, further ideas in the taking of penalties in soccer, and in stock control for perishable goods have been added. Readers seeking Solutions to the exercises should now contact Springer directly.

Brighton, UK
September 2019

John Haigh

Preface to the First Edition

The motivation for this book was to construct a mathematics module that encouraged First-Year university students to appreciate the diverse ways in which the mathematics they already knew, or were just about to learn, can impact on everyday occurrences. The applications discussed here will give students further practice in working with calculus, linear algebra, geometry, trigonometry and probability, thus supporting other modules. The chapters are largely independent of each other: and within most chapters, sections are often self-contained, giving teachers freedom in selecting topics to their taste. All in all, there is likely to be more material than would be covered in a one-semester course. Readers are assumed to have a background roughly equivalent to single-subject Mathematics at A-level in the UK.

In many degree courses, students have the opportunity, or even the obligation, to write an extended essay on a mathematical topic of their choice. They will find a range of ideas here, in the set exercises as well as the formal text, and references to more substantial accounts.

Mathematics has many applications: it is unrealistic to expect one book to do more than give a representative sample of them. My selection runs from the obviously practical, such as how to calculate mortgage repayments, or schedule nurses to cover hospital wards, to amusing ways in which mathematical ideas can give pointers to tactics in TV Game Shows, or influence the scoring systems in sporting events. We see how simple differential equations can model bacterial growth, mixing liquids, emptying baths, evaporating mothballs and the spread of epidemics; when looking at darts, roulette and how people progress through hierarchical organisations, we use combinatorics, logic, difference equations and calculations with matrices and vectors. The variety of voting systems in use throughout the world is inevitable, given Kenneth Arrow's 'Impossibility Theorems'—that no system can be constructed that satisfies *all* among a simple list of desirable properties! Mathematics is versatile: and discovering the unusual places where it provides insights or answers is enjoyable.

When young children first use the rule for dividing by a fraction—'turn it upside down and multiply'—they seldom understand why it works. But they quickly find that the rule is easy to apply, and gives the right answer. Only much later are they able to appreciate a valid justification. Similarly with more advanced material: heuristic explanations, or worked examples, can be more enlightening than an all-encompassing general approach. Meeting and understanding a formal proof of a theorem or technique the first time you use it is not essential. But it is vital that students appreciate the necessity for formal proofs at some stage of their studies, else they will not be confident about the conditions under which particular techniques can properly be applied. One of the most vital, if unacknowledged, words in mathematics is 'if': *if* certain conditions hold, *then* some result follows. All too often, it is tempting to use that result without checking the 'if' part.

Each chapter ends with a collection of exercises, which are an integral part of the book. Seeking to solve them should help students reinforce their understanding of basic principles. I do not indicate whether an exercise is expected to be routine, or quite tricky; lack of that knowledge is exactly the position mathematicians find themselves in when confronted with a problem to solve. I am happy to supply a set of Solutions to these exercises to any teacher who emails me at J.Haigh@sussex.ac.uk. (Students, please don't pretend. The best way to learn maths is to try to solve problems from scratch, without having looked at a solution first. And if handing in my work in your name would gain you significant credit, is your degree qualification worth having?)

It is a pleasure to acknowledge the feedback from the cohorts of Sussex students who have taken a module based on much of this material, as well as the helpful comments from three anonymous referees, and the support from the Springer production team.

The Appendix contains several formulae, techniques and approximations that working mathematicians will have used so often that they are as familiar as common multiplication tables. No student will regret the time spent in committing this list to memory.

Brighton, UK   John Haigh
September 2015

Contents

Chapter 1
Money

1.1 Interest

When you borrow or lend money for a period of time, 'interest' will be either charged or received. Without further elaboration, it might be *simple* or *compound* interest; to appreciate the difference, imagine investing £100 for 100 years at an annual rate of 5%.

(a) With simple interest, £5 is added each year, £500 in total, hence the end total is £600.

(b) With compound interest, let S and E, respectively, denote the amounts at the Start and End of a given year. During the year, 5% of S is added, so $E = S(1 + 5/100)$. This happens every year; thus, after 100 years the total amount we have is $£100(1 + 5/100)^{100} = £13, 150.13$.

From such dramatic differences, Albert Einstein concluded that compound interest is 'the greatest mathematical discovery of all time'. Unless otherwise specified, *all our calculations assume that compound interest is being used.*

Suppose you borrow an amount C, on which interest at an annual rate of $\alpha\%$ is charged. If you make no repayments, you will owe $C(1 + \alpha/100)^n$ after n years, so your debt will *double* at that time n when

$$C(1 + \alpha/100)^n = 2C,$$

i.e. when $n \log(1 + \alpha/100) = \log(2) = 0.693147181\ldots$. Thus

$$n = \frac{0.693\ldots}{\log(1 + x)},$$

where $x = \alpha/100$. Now when $|x| < 1$, $\log(1 + x) = x - x^2/2 + x^3/3 - \cdots$, so when x (> 0) is small, the right side is *just under* x, meaning that n is *just over* $0.693/x = 69.3/\alpha$. This background justifies

© Springer Nature Switzerland AG 2019
J. Haigh, *Mathematics in Everyday Life*,
https://doi.org/10.1007/978-3-030-33087-3_1

The Rule of 72 At an annual interest rate of $\alpha\%$, it will take approximately $72/\alpha$ years for a debt to double, if no repayments are made.

A similar argument, of course, shows that if you invest a sum of money that earns an annual interest of $\alpha\%$, it will take around $72/\alpha$ years for your capital to double. And if inflation runs at 3% annually, then it will take about 24 years to halve the value of any cash you keep under the bed.

There are two good reasons to replace the figure of 69.3 by 72. The first is that 72 is, helpfully, 'just over' 69.3; the other is that 72 has many divisors, often leading to easy arithmetic. For example, the Rule indicates that for interest rates of 6 and 12%, it will take about 12 years or six years, respectively, for a debt to double; exact calculations lead to 11.896 years and 6.12 years. This Rule is very handy.

Notice that the Rule gives an answer that is too large when the rate is 6%, too small when the rate is 12%. The Exercises ask you to *prove* these exemplify the general result that there is some interest rate (just under 8%), below which the Rule *overestimates* the doubling time, and above which the Rule gives an *underestimate*.

If the (annual) interest rate is 12%, and you borrow £1000 for six months, how much should you repay? It is tempting to think that, borrowing for half a year, the interest charge is half of 12%, i.e. 6%, so the repayment should be £1060, but this is incorrect. Let the interest rate for six months be $x\%$; after six months you owe $£1000\left(1 + \frac{x}{100}\right)$, so after another six months, one year in total, you would owe $£1000\left(1 + \frac{x}{100}\right)^2$. And since the annual rate is 12%, we have

$$1000\left(1 + \frac{x}{100}\right)^2 = 1000\left(1 + \frac{12}{100}\right)$$

leading to

$$1 + \frac{x}{100} = \sqrt{1.12} = 1.0583\ldots.$$

The interest rate for six months is just 5.83%, you should repay £1058.30.

Similarly, if the annual rate is 12% and you borrow £1000 for just one month, the appropriate interest rate is that value x so that

$$1000\left(1 + \frac{x}{100}\right)^{12} = 1000\left(1 + \frac{12}{100}\right),$$

giving $1 + \frac{x}{100} = 1.12^{1/12} = 1.0094888$ or so. The interest rate for one month is 0.94888%, you would repay £1009.49.

You may read 'The *quarterly* interest rate is 2%'. After a full year of four quarters, with 2% interest charged each time, a loan of size C becomes $C(1.02)^4 = 1.0824C$. The true annual interest rate is some 8.24%, not the 8% that the headline figure suggests. A rate of 1% per month would mean that the debt is $C(1.01)^{12} = 1.1268C$ at the year's end, a true annual rate of 12.68%, not 12%.

However the interest rate is presented, there is a simple way to find how much is owed at any specific future time: if the rate is not already given as the true annual rate, first convert it to that format; then, with $\alpha\%$ as the true annual rate, the amount owed when borrowing C for t years, whether t is an integer, a fraction, or whatever, is simply

$$C\left(1 + \frac{\alpha}{100}\right)^t.$$

We say that interest is *compounded continuously.*

Example 1.1 If the true (annual) interest rate is 10%, what should you pay back if you borrow £1000 for (a) six months; (b) 40 months?

Solution. (a) Six months is half a year, so the debt rises to £1000$(1.10)^{1/2}$, or £1048.81. (b) 40 months is three and a third years, so the debt would be £1000$(1.10)^{10/3} = $ £1373.96.

1.2   Present Value and APR

Would you prefer a gift of £1000 now, or a gift of £10,000 two years from now? Only rarely would the immediate money be chosen, but what if the alternative is £1100 in two years' time? The choice is no longer obvious. In general, we seek to answer the question: how much is it worth *now* to be offered amount X, but to be received in K years' time?

Suppose I is the current interest rate: if you had amount Y now, that would become $Y(1 + I)^K$ in K years, so setting $X = Y(1 + I)^K$ leads to $Y = X/(1 + I)^K$. Having $X/(1 + I)^K$ *now* is equivalent to having X in K years time: we say that the *Present Value*, or PV, of the offer is $X/(1 + I)^K$.

This idea lets us make sensible comparisons. In the scenario above, the Present Value of receiving £1100 two years hence would be equivalent to £1000 now, provided that, with an interest rate of I, $1100/(1 + I)^2 = 1000$, i.e. when $I = 4.88\%$. For lower interest rates, the delayed gift is preferred, for higher rates the immediate cash is more attractive. If a stream of money is expected at different future times, the PV of that stream is found by summing the PVs of each element. For example, a company might invest in an expensive piece of machinery now, with the prospect of a stream of higher future profits. The *times* at which those profits appear, as well as the amounts, are vital: an extra million pounds tomorrow is worth far more than a million pounds twenty years hence. Formally, if I is the interest rate, and amount c_j arises at the end year j for $j = 1, 2, \ldots, n$, then the PV of this income stream is $\sum_{j=1}^{n} c_j/(1 + I)^j$.

In the UK, there is a proposal to build HS2—a high-speed rail link from London via Birmingham to points further north, ready sometime after 2030. The cost has been estimated at over 50 billion pounds, and some economic justifications involve assessing the values of time saved over many years into the future. Simply adding up the sums involved, without taking account of WHEN costs will be incurred, and benefits seen, makes no sense.

Example 1.2 What (sensible) interest rates might make receiving £1000 *now* preferable to £2000 in one year's time, then repaying £910 one year later?

Solution. Work in £, with interest rate I. 'Repaying' an amount means 'receiving' the negative of that sum, so the Present Value of the second offer is $2000/(1 + I) - 910/(1 + I)^2$. The immediate gift is preferable if

$$1000 > \frac{2000}{1 + I} - \frac{910}{(1 + I)^2}.$$

Write $x = 1 + I$ and simplify: this is the same as $100x^2 - 200x + 91 > 0$. The quadratic factorises as $(10x - 13)(10x - 7)$, so its roots are at $x = 13/10$ or $x = 7/10$, corresponding to interest rates of $I = 0.3$ or $I = -0.3$. The quadratic is positive when $I > 0.3$ (and when $I < -0.3$).

Hence, if $I > 0.3$, i.e. interest rates exceed 30%, the PV of the immediate £1000 is higher, while for (positive) interest rates of below 30%, the delayed gift with repayment has higher PV.

You may be offered a loan of size L, to be repaid via instalments of sizes $S_1, S_2, S_3, \ldots$ at respective times $t_1, t_2, t_3, \ldots$ later. Another organisation may offer you the same amount, but with a different repayment schedule. To assess the merits of such offers, you could calculate their *Annual Percentage Rate*, or APR: this is defined to be that interest rate I that makes the Present Value of the repayments equal to the sum loaned.

To find the APR, calculate how much is still owing after each repayment: after repaying S_1 at time t_1, you owe $L(1 + I)^{t_1} - S_1$; at time $t_2 - t_1$ later, you pay S_2, hence you now owe

$$(L(1 + I)^{t_1} - S_1)(1 + I)^{t_2 - t_1} - S_2 = L(1 + I)^{t_2} - S_1(1 + I)^{t_2 - t_1} - S_2.$$

Continue this argument, to see that the amount left after the nth repayment collapses to

$$L(1 + I)^{t_n} - S_1(1 + I)^{t_n - t_1} - S_2(1 + I)^{t_n - t_2} - \cdots - S_n.$$

Equate this to zero to mark the final repayment, divide through by $(1 + I)^{t_n}$ to obtain the formula

$$L = \sum_k \frac{S_k}{(1 + I)^{t_k}}.$$

Solve this to find I. The APR is meant to reflect the true cost of a loan, taking into account any fees, the frequency and amounts of repayments, all calculated in the same 'fair' way.

Example 1.3 You wish to borrow £5000. Megaloan asks for repayments of £1500 each year for five years, Affordit will seek £1000 each year for nine years. Repayments begin 1 year after the loan. Which APR is lower?

Solution. For Megaloan, we solve $5000 = 1500 \sum_{k=1}^{5}(1/(1+I))^k$. Replacing $1/(1+I)$ by x, this is the same as $5000/1500 = 10/3 = x + x^2 + \cdots + x^5$. How best to solve this? It has the format $A = x + x^2 + \cdots + x^K$, i.e. $A = x(1 - x^K)/(1 - x)$, and so can be turned into '$x = g(x)$' as

$$x = (A + x^{K+1})/(A + 1).$$

The iteration scheme $x_{n+1} = g(x_n)$, with a suitable starting value such as $x_0 = 0.9$ can be used here to find x, and hence I; the APR is 15.24%. A similar argument for the terms offered by Affordit leads to

$$5000/1000 = 5 = x + x^2 + \cdots + x^9,$$

with an APR of 13.70%. Although you would repay £9000 to Affordit, but only £7500 to Megaloan, Affordit offer a lower APR.

1.3 Mortgage Repayments; Annuities

The largest financial commitment that many people make is to borrow money to buy a house. If you borrow amount C, at annual interest rate $I > 0$, paying off the same amount each month for n years, how much will you pay?

Consider a month at a time: let C_k be the amount owed at the end of month k, so that $C_0 = C$, and suppose that R is repaid each month, beginning one month after you receive the loan. With an annual interest rate of I, suppose the corresponding *monthly* rate is α, found via $1 + I = (1 + \alpha)^{12}$. At the end of the first month, the amount owing is $C_1 = C(1 + \alpha) - R$, (yes?).

At the end of month two, we similarly have $C_2 = C_1(1 + \alpha) - R$, so, using the previous expression, we obtain $C_2 = C(1 + \alpha)^2 - R(1 + \alpha) - R$.

Then, from $C_3 = C_2(1 + \alpha) - R$, we get

$$C_3 = C(1 + \alpha)^3 - R(1 + \alpha)^2 - R(1 + \alpha) - R,$$

from which the general relation

$$C_k = C(1 + \alpha)^k - R \sum_{j=0}^{k-1}(1 + \alpha)^j \tag{1.1}$$

is obvious. There are $m = 12n$ months in n years, so repaying the whole amount after m months means that $C_m = 0$; hence we find R from the equation

$$0 = C(1+\alpha)^m - R \sum_{j=0}^{m-1}(1+\alpha)^j.$$

The sum is a geometric series. Recall that, when $x \neq 1$,

$$\sum_{j=0}^{m-1} x^j = (x^m - 1)/(x - 1),$$

so, after a little manipulation, since $\alpha > 0$,

$$R = \frac{C\alpha(1+\alpha)^m}{(1+\alpha)^m - 1} \tag{1.2}$$

is the monthly repayment.

Since we found α via $(1 + I) = (1 + \alpha)^{12}$, then (1.2) can also be written as

$$R = \frac{C\alpha(1+I)^n}{(1+I)^n - 1}.$$

Example 1.4 You borrow £100,000 at 6% annual interest to be repaid over 25 years. (a) How much is repaid each month? (b) How much is still owing after 15 years? (c) Is it surprising that, after 15 years, you still owe more than half of what you borrowed, despite having made 60% of the payments due?

Solution. (a) We have $C = 100,000$, $I = 0.06$ and $n = 25$, so α comes from $1.06 = (1 + \alpha)^{12}$, i.e. $\alpha = 0.4867\%$. From Eq. (1.2), $R = 634.62$—you pay about £635 each month.

(b) After 15 years, i.e. 180 months, (1.1) shows that the amount still owing is

$$C_{180} = C(1+\alpha)^{180} - R\sum_{j=0}^{179}(1+\alpha)^j = C(1+\alpha)^{180} - \frac{R((1+\alpha)^{180} - 1)}{\alpha}.$$

Arithmetic gives an answer of 57,575.55. You still owe about £57,600.
(c) Each payment must pay off the interest accumulated since the last one, and also reduce the capital owed. In the early years, when the amount owed is greatest, the interest component is higher, so less goes towards reducing the amount still owed.

Example 1.5 An outfit is on sale at £200. If you take out a credit card to purchase it, with an annual interest rate of 30%, the store will reduce the price by 10%; you will pay this off, making equal instalments at the end of each month, for n months. In total cash terms, compare (a) paying the full price upfront, and (b) taking the store card, and paying the loan off over 8, 12 or 24 months.

Solution. Let R_m be the monthly repayment if you take the credit card loan of £180 for m months. An annual rate of 30% leads to a monthly rate of $x\%$, where

$(1 + x/100)^{12} = 1.30$, so that the monthly rate is $\alpha = 2.21\%$. Equation (1.2) shows that

$$R_m = (180 \times 0.0221 \times (1.0221)^m)/((1.0221)^m - 1),$$

and the total repaid is plainly $m.R_m$. Over 8 months, repayments are £24.79, so the total is £198.32; similarly, $R_{12} = £17.24$, giving a total of £206.88; and $R_{24} = £9.74$, or £233.76 overall. Using the store card to obtain the discount is cheaper than paying cash if you repay over 8 equal monthly instalments, but the cash amount is less than the credit cost if you repay over 12 or 24 months.

Many people take out an *annuity* to provide an income for their retirement. They hand over a lump sum to an insurance company, in exchange for receiving a fixed amount monthly for the rest of their life, no matter how long or how short. Exactly the same mathematics as is used to calculate mortgage repayments also applies here!

The reason is that receiving an annuity is the mirror image of paying off a mortgage. C is now the amount handed over, I is the annual interest the company expects to earn from its investments, over and above its own costs, and n is the number of years they expect to make payments, based on your age and health. With α again the monthly interest corresponding to I, the expression (1.1) for C_k now represents how much the insurance company retains after k payments to you. With $m = 12n$, setting C_m equal to zero means that, on average, the insurance company expects to return the capital it received.

Some annuitants will live longer than average, others will die earlier; the company relies on having a large number of customers so that their estimate of the average lifespan is accurate. The residue from those who die an early death will enable the company to meet the payments to people like Jeanne Louise Calment who, at the age of 90, promised her modest apartment to André-François Raffray, in exchange for a monthly sum of around $500 for the rest of her life; she died, 32 years later, having been paid the value of her apartment many times over.

To cater for the effects of inflation, an annuitant may ask that her annual income rises steadily. See the Exercises for the simple adjustment that is required to recalculate the new payments.

1.4 Investing

Conventional wisdom is that stocks and shares, although riskier, will tend to produce a higher return in the long term than deposits in bank accounts. The price of shares fluctuates, according to what 'the market' thinks they are worth at any time. Suppose that an investor, seeking to commit a total of $C > 0$ during a year, can buy shares either

(a) just once, at the average price over the year; or
(b) spending C/n on n occasions at the current price, which varies around that average.

Which method is expected to produce most shares? Or does it make no difference?
    Suppose $\{x_i : i = 1, 2, \ldots, n\}$ are the share prices at times $\{1, 2, \ldots, n\}$, all strictly positive, and that θ is a real number. Then as the sum of non-negative numbers is non-negative, we have

$$\sum_{i=1}^{n} \left(\theta \sqrt{x_i} + \frac{1}{\sqrt{x_i}} \right)^2 \geq 0,$$

i.e.

$$\theta^2 \sum_{i=1}^{n} x_i + 2n\theta + \sum_{i=1}^{n} \frac{1}{x_i} \geq 0.$$

But if a quadratic in θ is never negative, it cannot have two distinct real roots, so its discriminant is non-positive ('$b^2 \leq 4ac$'), i.e.

$$n^2 \leq (\sum_{i=1}^{n} x_i) \left(\sum_{i=1}^{n} \frac{1}{x_i} \right).$$

So if $\bar{x} = (\sum_{i=1}^{n} x_i)/n$ is the *arithmetic mean* of the x's, we see that

$$\frac{n}{\bar{x}} \leq \sum_{i=1}^{n} \frac{1}{x_i} \quad \text{and so} \quad \frac{C}{\bar{x}} \leq \sum_{i=1}^{n} \frac{C/n}{x_i}.$$

In this final inequality, the right side is the total number of shares that would be acquired when spending C/n on n occasions. And since $\bar{x}$ is the average price at those times over the year, so the left side is the total number of shares expected from spending C in one fell swoop. Thus:

Theorem 1.1 *However the share price fluctuates, we expect* at least as many *shares if we make n regular purchases, rather than by investing C all at once at their average price.*

    This principle is known as 'pound cost averaging'. The extra shares you obtain when the price is low *more than makeup* for the fewer you get when the price is high. And as the Exercises demonstrate, the more the share price fluctuates, the greater the advantage of drip-feeding your funds over the year.

Example 1.6 The prices (pence) of shares in Tescos on the first trading day of the month from June 2014 to May 2015 were: 302, 284, 258, 229, 186, 173, 185, 189,

227, 244, 240, 218. How many shares would you possess if you invested £12,000
(a) by spending £1,000 on the first trading day of each month; (b) by spending the
whole lot at the beginning of the period; (c) by spending the whole lot at the average
price over the year? (Ignore dealing costs.)

Solution. (a) We calculate $\sum_{r=1}^{12} \frac{1000}{x_r}$ where x_r is the price (pounds) in the rth month.
It works out as 5418.6 shares.
(b) This is simply $12,000/3.02 = 3973.5$ shares.
(c) In the notation of (a), the average price (in pounds) is $\sum_{r=1}^{12} x_r/12 = 2.279$, so
we would have $12,000/2.279 = 5265.1$ shares.

So *pound cost averaging*, as compared to a single purchase at the annual average
price, gives $5418.6 - 5265.1 = 155.5$ more shares, an increase of $155.5/5265.1 \times 100\%$, i.e. nearly 3%.

1.5 Personal Finance

Savings Even those who will qualify for a maximum State pension will usually wish
to make extra provision from savings built up during their working life. So it is worth
making calculations, however speculative, about how much you need to save in order
to build up a suitable sum. At its simplest, if you invest X at the beginning of each
year for n years, and the annual growth rate is $G > 0$, the mathematics is similar to
that used for mortgage repayments.

In this context, write Y_k as the amount you will have at the end of year k, so that
$Y_0 = 0$, $Y_1 = X(1 + G)$ (we invest at the beginning of the year), $Y_2 = Y_1(1 + G) + X(1 + G) = X(1 + G)^2 + X(1 + G)$, etc. giving

$$Y_n = X \sum_{j=1}^{n} (1 + G)^j = \frac{X(1 + G)((1 + G)^n - 1)}{G}.$$

Thus, for example, compare twins Anne, who invests X each year for 40 years, and
Beth who starts 20 years later, and invests W each year for just 20 years: how much
more must she save to achieve the same final amount as her sister, if both enjoy the
same growth rate?

For Beth to do as well as Anne, she requires that

$$\frac{X(1 + G)((1 + G)^{40} - 1)}{G} = \frac{W(1 + G)((1 + G)^{20} - 1)}{G},$$

i.e. that $W = X((1 + G)^{20} + 1)$. At a modest growth rate of 3%, Beth must save
almost three times as much as Anne every year, and if the growth rate is 6%, she
must save 4.2 times as much. The magic of compound interest demonstrates the
benefits of saving from a young age.

In general, saving X every year for n years will give the same total as saving W every year for m years, whenever

$$\frac{X(1+G)((1+G)^n - 1)}{G} = \frac{W(1+G)((1+G)^m - 1)}{G},$$

or

$$\frac{W}{X} = \frac{(1+G)^n - 1}{(1+G)^m - 1}.$$

Let's assess how the growth rate affects the value of this ratio for general values of m and n. The simplest case should be when $G = 0$, but putting $G = 0$ in this formula leads to the nonsensical expression $0/0$—(maths students who write that ratio down, without comment, risk the wrath of their tutors). We need *L'Hôpital's Rule*—see the Appendix.

This Rule says that, at some point c where $f(c) = g(c) = 0$, then, provided all expressions make sense, the limit of $f(G)/g(G)$, as $G \to c$, is $f'(c)/g'(c)$. Here take

$$f(G) = (1+G)^n - 1, \quad g(G) = (1+G)^m - 1 \quad \text{and } c = 0.$$

Plainly $f'(G) = n(1+G)^{n-1}$, $g'(G) = m(1+G)^{m-1}$, so their ratio $f'(G)/g'(G)$ reduces to $(n/m)(1+G)^{n-m}$. L'Hôpital's Rule tells us that, as $G \to 0$, so the ratio W/X tends to n/m—which makes perfect sense: with zero growth, then if you invest for m years rather than n, the amount required to give the same total is just n/m times as much each year.

Good mathematicians get in the habit of making such simple checks.

Tax In the UK, individuals under State Pension Age with income below £100,000 are taxed on their income from employment and pensions as follows. There is a Personal Allowance, P, on which no Income Tax (IT) is levied; above this, IT is levied at the Basic Rate of 20% on the amounts from P to $P + B$ for some $B > 0$, and at 40% on amounts above $P + B$. P and B are updated annually.

National Insurance Contributions (NICs) are levied on *earned* income (but not on pensions or annuities); no NICs are paid on earnings up to a Lower Limit L, 12% is charged on earnings between L and the Upper Limit U, and 2% on earnings above U. Again, L and U change annually.

$T(x)$, the *marginal rate of tax* at a total income level of £x (as a percentage), is defined as the extra tax due, from both IT and NIC (in pence), if income rises from £x to £$(x + 1)$. The total 'tax' paid by someone whose income is X is

$$\sum_{x=0}^{x=X-1} T(x);$$

the derivation of an integral as the limit of a sum shows that this is close to $\int_0^X T(x)\,dx$.

'Tax years' start on 6 April. In England in 2019/20, $P = £12{,}500$, $B = £37{,}500$, $L = £8632$ and $U = £50{,}024$.

Taking an Annuity The principles behind the next example may be relevant to some of your family now, or to you at some stage of your life.

Example 1.7 David, aged 68, will use his savings to buy an annuity. If he buys one now, and continues to work, his salary is large enough that the annuity will be taxed at 40%; but if he postpones buying it until he stops work in five years time, it will be taxed at just 20%. Postponing taking the annuity also means that the amount it pays will increase by some factor $K > 0$. If his only consideration is financial, how should David decide between those alternatives?

Solution. One sensible path is to compare their Present Values. Assume his future lifespan is n years, with $n > 5$, write $A = $ Annual amount from the annuity, (assumed paid at the beginning of a year), and take I as the Interest rate. We calculate the PVs of the total amounts David receives, after tax.

The PV of the annuity if he draws it immediately is

$$\sum_{r=1}^{5} \frac{(0.6)A}{(1+I)^{r-1}} + \sum_{r=6}^{n} \frac{(0.8)A}{(1+I)^{r-1}}.$$

Postponing for five years, the PV is

$$\sum_{r=6}^{n} \frac{(0.8)A(1+K)}{(1+I)^{r-1}}.$$

Thus drawing now is better than postponing if, and only if, the first expression exceeds the second. This simplifies to

$$3\sum_{r=1}^{5} \frac{1}{(1+I)^{r-1}} > 4K \sum_{r=6}^{n} \frac{1}{(1+I)^{r-1}}.$$

Write $x = 1/(1+I) < 1$; we are summing geometric series, so the condition reduces to

$$3(1 - x^5) > 4Kx^5(1 - x^{n-5}).$$

As we expect, the values of n for which this holds depend on x and K.

A realistic value of K is about 35% (based on the annuity rates currently available, and assuming his savings pot increases modestly over the five years if he postpones buying the annuity); suppose we take $I = 4\%$, so that $x = 1/1.04$. With those figures, the criterion becomes $0.5342 > 1.1507(1 - x^{n-5})$, or, using logs, $n < 20.9$. Thus, on those figures, he should take the annuity immediately if he expects to die within about 21 years, but if he expects to live longer, he should postpone for five years.

But suppose I is different: with $I = 2\%$, the cut-off figure is about 18 years, while with $I = 6\%$, it increases to about 27 years. How you value prospective income in the far future affects the decision you make now. And other assumptions about K, or changes in tax rates, may also lead to different conclusions.

The Exercises examine how this analysis is easily modified if the annuity increases by the factor $1 + \alpha$ each year. The actual annuity amount in year r changes from A to $A(1 + \alpha)^{r-1}$, so, if you look closely, you will see that we would now write $x = (1 + \alpha)/(1 + I)$, and follow the same path!

A similar dilemma is addressed by Dagpunar (2015): if UK pensioners postpone taking their State Pension, the actual amount to be paid is enhanced (by 5.8% for each year of delay from 2016). By using tables of life expectancy at various ages, Dagpunar compares the total amounts pensioners can expect to receive with or without postponement; his conclusion is that unless there is a good reason to believe your life expectancy is well above the UK norm, deferring is probably not worthwhile.

(He carefully notes that his article should not be construed as offering financial advice, or advocating any particular course of action. I make a similar disclaimer about the contents of this book.)

Student Loans For UK Student Loans beginning in 2012 or later, repayments may be due from the April after graduation. In each year from then, 9% of total income in excess of some *Threshold*, T, is paid until the debt is cleared, or 30 years elapse, whichever comes earlier. No payments are required in years when income is below T. Interest is charged at the rate of inflation, plus an amount that depends on the income that year: this extra amount is zero for incomes below T, it is 3% for incomes above some higher level E, and on a linear sliding scale for incomes between T and E. In 2018/19, $T = £25,000$ and $E = £45,000$; these figures are scheduled to rise in line with changes in average earnings.

Example 1.8 With the information above, and assuming the values of the threshold T and the higher amount E remain fixed over the next few years, draw up a table showing how the Student Loan for Cindy, who has an initial Loan of £30,000 and a starting salary of £25,000, changes. Assume her salary increases by £2,000 each year, that the inflation rate is 2%, and (for simplicity) that she receives her entire salary at the beginning of the year, but pays the interest due at the year's end.

Solution. When her salary is £S, with $25,000 \leq S \leq 45,000$, the interest rate is

$$(2 + 3(S - 25,000)/20,000)\%.$$

A suitable table looks as follows: (Money is in pounds, and the column 'End Loan' means after interest added and repayment made. All calculations rounded to the nearest pound.)

Year	Salary	Interest Rate (%)	Start Loan	Interest	Repay	End Loan
1	25,000	2.00	30,000	600	0	30,600
2	27,000	2.30	30,600	704	180	31,124
3	29,000	2.60	31,124	809	360	31,573
4	31,000	2.90	31,573	916	540	31,949
5	33,000	3.20	31,949	1022	720	32,251

Despite 'repaying' £1800 in total, she owes £2251 *more* than when she started. And even after five years, her repayments still do not match the interest charged.

1.6 More Worked Examples

Example 1.9 Suppose you invest £10,000 in property, which increases in value at the compound rate of 5% annually, but you are charged 2% of the total value at the end of each year. What will your investment be worth at the end of five years?

Solution. If, at the beginning of any year, your investment is worth C, it increases to $1.05C$, but the 2% charge then turns this into $1.05C \times 0.98 = 1.029C$, so at the end of five years, you will have $1.029^5 C$. Here this translates to £11, 536.57 or so. (Note that a 5% increase, with a 2% decrease, is not the same as a 3% increase.)

Example 1.10 On your retirement, you have a lump sum of £200,000 to provide an income until you die. An insurance company estimates that you will live for 20 years, and assumes it can earn 4% (after its expenses) on the capital you give it. What (fixed) amount would you expect to receive each year (round to a multiple of £100)?

You believe that you can earn 5% yourself through your skilled investing: suppose you are correct, and that you take a fixed income of £20,000 at the end of each year. How long will your pot last?

Solution. For the company payment, we use (1.2), with $\alpha = 4\%$ and $m = 20$. Working in pounds, you would expect

$$200,000 \times 0.04 \times 1.04^{20}/(1.04^{20} - 1) = 14,716.35,$$

which rounds to £14,700.

In your own hands, let C_k be your capital at the end of year k; then, so long as your plan lasts, we have $C_{k+1} = 1.05C_k - S$, where S is the income you draw. Thus

$$C_k = 1.05^k C_0 - S(1 + 1.05 + \cdots + 1.05^{k-1}) = 1.05^k C_0 - S\frac{1.05^k - 1}{1.05 - 1}.$$

With $C_0 = 200,000$ and $S = 20,000$, and so long as $C_k > 0$, this reduces nicely to $C_k = 200,000(2 - 1.05^k)$. So we need $1.05^k < 2$.

Now $1.05^{14} = 1.9799\ldots$, $1.05^{15} = 2.0789\ldots$. You can live this lifestyle for 14 years, and since $C_{14} = 4013.68$, your pot has diminished to about £4000 at the beginning of year 15.

Example 1.11 To borrow £100, a payday loan company will charge 0.8% per day *simple* interest. How much will you pay back if you clear your debt after 13 days? And what is the APR?

Solution. The total interest charge is £$0.80 \times 13 = $£10.40, so you must pay £110.40 to clear the debt. If the APR is α, we must solve

$$100 \left(1 + \frac{\alpha}{100}\right)^t = 110.40,$$

where $t = 13/365$ is the fraction of a year for which you borrowed the money. Thus

$$\frac{13}{365} \log \left(1 + \frac{\alpha}{100}\right) = \log \left(\frac{110.40}{100}\right) = \log(1.104) = 0.09894,$$

i.e. $\log(1 + \frac{\alpha}{100}) = 2.7779$, so $\frac{\alpha}{100} = 15.09$, meaning that the APR to be quoted is, in round figures, 1509%.

Example 1.12 At the age of 25, you decide to build up a pot of money to fund your retirement. You assess that you will retire at age 70, and will need a lump sum of one million pounds at that time. You will pay into the pot at the end of each year, you expect to increase the amount you pay by 2% each year, and that your funds will grow at 4% annually. How much should you plan to save in the first year? If you were to put off starting to save for ten years, what would be the new initial sum? (Assume your annual savings are paid in at the end of each year.)

Solution. Suppose you have C_n at the beginning of year n, and you start by saving R in the first year. Then $C_0 = 0$ and $C_{n+1} = C_n.1.04 + R.1.02^{n-1}$, so $C_1 = R$, $C_2 = (1.04 + 1.02)R$, $C_3 = (1.04^2 + 1.04 \times 1.02 + 1.02^2)R$, which leads to

$$C_n = R \sum_{i=0}^{n-1} 1.04^i \times 1.02^{n-1-i} = R \frac{1.04^n - 1.02^n}{1.04 - 1.02} = 50R(1.04^n - 1.02^n)$$

(summing the geometric series). In the first case, saving for 45 years, we want

$$50R(1.04^{45} - 1.02^{45}) = 1,000,000,$$

i.e. $R = 20,000/(1.04^{45} - 1.02^{45}) = 5876.61$. You need to save about £5,900 in the first year.

In the second case, replacing 45 by 35 means that $R = 10,276.44$; you must now save about £10,300 in the first year. Be warned!

Example 1.13 A student graduates with a Student Debt of £30,000, and her share of the monthly rental on her London flat will be £1000 (including bills). She desires a monthly income of at least £800, after all taxes, debt repayments, and bills. Suppose the Personal Allowance for Income Tax is $B = £12,500$ and take the Lower threshold for National Insurance Contributions as $L = £8,600$, for simplicity. Tax rates are as given in the text. Find the minimum annual salary she needs. If she were paid that salary (rounded up to the next £100), how much would she keep of the 10% bonus she gets?

Solution. She hopes to have at least £1,800 each month after all taxes and debt repayments, i.e. at least £21,600 per year. So let's confidently assume her gross salary needs to be at least £25,000 (where Student Loan payments kick in). On a gross salary of exactly £25,000, she loses $£12,500/5 = £2500$ in IT and $£16,400 \times 0.12 = £1968$ in NIC, so she retains just £20,532. She now loses 41 pence in each pound over 25,000, so to get her extra £1068 to bring her total to £21,600 she needs another $£1068/0.59 = £1810.17$. Rounding, her salary must be at least £26,800.

If she received a 10% bonus, i.e. £2680, she keeps only 59% of this (it is still subject to IT, NIC and Loan), so she keeps £1581.20.

1.7 Exercises

1.1 How good is the 'Rule of 72' for interest rates of 4% and of 24% (i.e. percentage-wise, how close to the exact value does this Rule come)?

1.2 Let $\alpha\%$ denote the interest rate. *Prove* that there is some value x such that this Rule *overestimates* the doubling time when $0 < \alpha < x$, and *underestimates* it if $\alpha > x$. Find the value of x (correct to 3 significant figures).

1.3 Suppose a loan is offered, on the terms that interest will be '15%, compounded quarterly'. Show that the true annual rate is about 15.865%.
Find the true annual percentage rate, correct to 3 decimal places, if the nominal rate is 15% compounded (i) monthly (ii) weekly (iii) daily (iv) hourly (v) even more frequently!

1.4 Listed are three alternative income streams: the sums shown are the amounts (in £1000) that will be paid to you at the ends of years 1, 2, 3, 4, 5.

A: 12, 14, 16, 18, 20 (total 80)
B: 16, 16, 15, 15, 15 (total 77)
C: 20, 16, 14, 12, 10 (total 72)

Consider possible interest rates of 10, 20 and 30%. Find the present values of each income stream, and decide, for each interest rate, which stream has the largest present value.

1.5 Peter asks to borrow £100 from you, and will pay this off by two payments of £70 at the end of years One and Two. Use the notion of Present Value to determine the interest rate to which this is equivalent.

1.6 Shares in Butterbean plc will pay annual dividends, of initial amount D one year after purchase. This amount will increase at the compound rate α each year (with $\alpha < I$), I being the relevant interest rate for computing Present Values. Show that the Present Value of the total of all the dividends that will ever be paid into the indefinite future is $D/(I - \alpha)$.

1.7 Determine the APR, correct to three significant figures, when a loan of £1000 now is to be paid off by five equal annual instalments of £250, starting one year from now.

1.8 As in Example 1.11 above, you arrange to borrow £100 at *simple* interest of 0.8% per day but now for (a) 8 days (b) 20 days. Find the respective APRs.

1.9 To buy a house, you borrow £100,000, to be repaid in equal *monthly* instalments for the next 25 years.

  (a) Find the repayments at annual interest rates of (i) 5% and (ii) 10%.
  (b) In total, how much more does the second case cost than the first?
  (c) In each case, how much do you owe after ten years of repayments?
  (d) *Sketch* a graph showing the amount owing against time when the interest rate is 10%.

1.10 Suppose you take out a mortgage of size $C > 0$, which you will repay by monthly instalments of some fixed size R. The *monthly* interest rate is $\alpha > 0$, and the first repayment is made one month after the loan is taken out, and the loan will be fully repaid after N months. Write $R = KC\alpha$ for some $K > 1$. Deduce that

$$N = \frac{\log(K/(K - 1))}{\log(1 + \alpha)},$$

and use this result to show that, when α is small, and you make fixed monthly repayments of size $2C\alpha$, you will repay approximately $1.4C$ altogether, over the lifetime of the mortgage.

1.11 We have seen that the formula for an amount of a level annuity is the same as that used for mortgage repayments. Suppose the insurance company will earn interest at rate $\gamma > 0$ per year, while you want the annuity to increase at rate $\beta > 0$ each year. Adapt the argument leading to (1.2) to cover this case, so finding the amount of the (initial) annual annuity payment, assuming $\gamma \neq \beta$. By taking γ fixed, and allowing β to approach γ, use L'Hôpital's Rule to find the initial annuity amount when $\gamma = \beta$.

1.12 At the age of 25, Colin decides to save a proportion of 10% of his income annually. His twin brother David will delay starting saving for k years, and then will also save a fixed proportion of his income; both will retire at age 68. Their incomes are identical, and rise by 2% each year. For a growth rate of 4%, what proportion of his income must David save, to put himself in the same position as Colin at retirement age, when (i) $k = 5$, and (ii) $k = 10$.

1.13 Your credit card company charges interest monthly, at the annual rate of 25%, on any outstanding debt after the monthly repayment. The rules on this monthly repayment say that any sum of up to £5 is to be paid off in full, otherwise you must pay at least £5, or all the outstanding interest plus 1% of the balance, whichever is greater. Making the minimum repayment, how long would it take you to reduce a debt of £5000 to at most £100?

1.14 (a) Over twelve consecutive months in 2014/15, the costs of units in the Neptune Russia and Greater Russia fund (in pence) were 249, 234, 268, 275, 251, 259, 237, 232, 222, 175, 169, 201; their average is 231. If you had £12,000 to spend, how many units would you have if you spent all of it at that average price? How many would you have if you spent £1,000 each month, at that month's price?

(b) If, in the 6 months where the price was below 235 pence, the prices had been 30 pence lower, while in the other 6 months, they had been 30 pence higher, what would the new figures be for those calculations?

(c) Comment on how the advantage of pound cost averaging appears to vary with the *variability* of the share price.

1.15 Suppose that, in a certain tax year, the values of P, B, L and U are as given in the text, and let $T(x)$ be the marginal tax rate at income level x.

(i) For someone whose income is x, all of it from employment, plot the graph of the function $T(x)$ on the range $0 < x < £72,000$.

(ii) Do the same for someone who gets a pension of £20,000, but also has other earnings, up to £52,000. (Treat the pension income as the first £20,000 of income, with the 'other earnings' coming after this.)

(iii) For each of (i) and (ii), is $T(x)$ monotone increasing?

1.16 Alice is aged 60, enjoys her job, expects to retire in ten years time, and has a savings pot to use to buy an annuity sometime. While she works, her annuity income would be taxed at 40%, in retirement it will be taxed at just 20%. An annuity would pay 5.1% of her savings pot at age 60, 6.8% at age 70, and she estimates her savings pot will increase annually at 4% until she starts the annuity if she postpones. Advise Alice on how long she would need to live, in order to justify postponing for 10 years, using the Present Value of the total sum, after tax, that she would receive from her annuity as the criterion. Assume she receives the annual payments at the beginning of each year, with an interest rate of 5% to give Present Values.

1.17 Use the text figures for P, B, L and U for the following circumstances: Lucy, aged 22, has just graduated, and has a job offer at an annual salary of £S. She has no other form of income. In addition to IT and NIC, she pays 9% of any excess salary over £25,000 for her Student Loan (SL). How much (to the nearest £100) must her salary be so that, after all deductions, she has a net *monthly* income of at least (i) £2000 (ii) £2500 (iii) £3000?

[It is clear that S must exceed 25,000 in every case, so follow the path of Example 1.13: assess how much she keeps of the first £25,000, then, taking account of all three deductions, see how much extra she needs to reach the stated net levels.] (iv) Suppose her starting salary is £26,000, and her initial SL is £25,000. Assume that the inflation rate is constant at 2%, the threshold for SL repayments remains fixed at £25,000, her salary is paid at the start of the year and increases by £1,000 annually; she pays the amounts due immediately. Interest on the SL is the inflation rate $+3(S-25,000)/20,000\%$ when $25,000 \le S \le 45,000$. Construct a table showing, for five years, her salary, the relevant interest rate that year, the SL amount owed at the start of the year, the interest charged, the repayment, and the amount owed at the end of the year. Show that she would owe over £26,800 at the start of the sixth year. (Work in whole pounds.)

Reference

Dagpunar J (2015) Deferring a state pension - is it worthwhile? *Significance* 12(2) 30–35

Chapter 2
Differential Equations

2.1 What They Are, How They Arise

Many problems in maths are of the form: 'We have two quantities of interest, and some information on how they are related. How to use that information to say as much as possible, and as precisely as possible, what that relation is'. A *differential equation* may arise when the relation between these quantities includes information about the *rate* at which one variable changes as the other one changes.

For example, let t represent time and let y be the distance travelled as in Fig. 2.1. If, as t increases from t_1 to t_2, so y increases from y_1 to y_2, then the *average velocity* over that period is $(y_2 - y_1)/(t_2 - t_1)$, the slope of the line AB.

But what about the actual velocity at time t_1? Keeping t_1 fixed, move t_2 closer and closer to t_1; write $t_2 = t_1 + \delta t$, where δt is a tiny quantity, and suppose the corresponding distance moved is $y_2 = y_1 + \delta y$. Then the average velocity over the time from t_1 to $t_1 + \delta t$ is the ratio $\frac{\delta y}{\delta t}$. In the limit, as δt decreases down to zero, this is the actual velocity at the time t_1, and we write it $\frac{dy}{dt}$. It might be more meaningful to write it as

$$\frac{d}{dt}(y),$$

and read this as 'the rate of change of distance y with time t', but we will stick to $\frac{dy}{dt}$. This will be the slope of the tangent to the curve at the point A.

Since $\frac{dy}{dt}$ represents velocity, so

$$\frac{d}{dt}\left(\frac{dy}{dt}\right) = \frac{d^2 y}{dt^2}$$

is the rate of change of velocity with time, i.e. the *acceleration*. If a stone of mass m (>0) travels vertically (and we can ignore air resistance), then Newton's Law, that Mass $\times$ Acceleration $=$ Force, says that

© Springer Nature Switzerland AG 2019
J. Haigh, *Mathematics in Everyday Life*,
https://doi.org/10.1007/978-3-030-33087-3_2

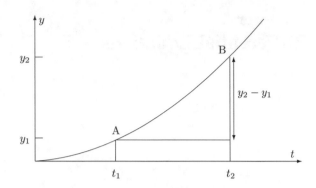

Fig. 2.1 Average velocity over a period of time

$$m\frac{d^2y}{dt^2} = mg, \quad \text{or simply} \quad \frac{d^2y}{dt^2} = g, \tag{2.1}$$

where g is the acceleration due to gravity. Our task would be to derive a formula for the distance y fallen at time t.

If air resistance is proportional to velocity, this changes to

$$\frac{d^2y}{dt^2} = g - c\frac{dy}{dt} \tag{2.2}$$

for some constant $c > 0$. Notice that, if we take $v = \frac{dy}{dt}$, this same equation becomes

$$\frac{dv}{dt} = g - cv. \tag{2.3}$$

For another example, suppose bacteria grow freely in a nutrient solution, with the population growth rate proportional to the current size. Here, with the natural notation that t is time, and x is population size, allowed to assume continuous values, we have

$$\frac{dx}{dt} = Kx, \tag{2.4}$$

$K > 0$ some constant. On the other hand, if there is competition for space or food, with a maximum possible size of N, the growth rate may be similar to (2.4) when x is small, but reduce as x increases: perhaps

$$\frac{dx}{dt} = Kx\left(1 - \frac{x}{N}\right) \tag{2.5}$$

is more realistic. Or maybe, as well as competition, the growth rate naturally decreases with time: perhaps K gets replaced by $K_0 \exp(-\lambda t)$. Then our relation becomes

$$\frac{dx}{dt} = K_0 \exp(-\lambda t)x\left(1 - \frac{x}{N}\right). \tag{2.6}$$

The general idea is to introduce symbols to represent relevant quantities, and then turn the physical description of what we believe is happening into an equation, or set of equations, intended to describe that belief.

Example 2.1 A snowplough, which clears snow at constant rate V/unit time sets off at time $t = 0$, and travels distance x_t by time $t > 0$. Snow began falling earlier, at time $t = -T$, and falls at a constant rate of s/unit time. Over the short time interval from time t to time $t + \delta t$, the amount of snow cleared is plainly $V \delta t$. But we can obtain another expression: as snow has been falling at rate s for a total time of $T + t$, the total amount of snow that fell at any position is $s(T + t)$; and as the plough moves distance $x_{t+\delta t} - x_t$ over that short time interval, the amount cleared can also be written as $s(T + t)(x_{t+\delta t} - x_t)$.

Equating these, we have

$$s(T + t)(x_{t+\delta t} - x_t) = V \delta t.$$

Divide both sides by δt, pass to the limit as $\delta t \to 0$ to obtain

$$s(T + t)\frac{dx}{dt} = V.$$

Example 2.2 Consider a bath or sink with a small plughole with cross-sectional area A. At time t, take $V = V_t$ as the volume of water, x as the depth and u as the escape velocity down the plughole. *Torricelli's Law* states that $u = \sqrt{2gx}$ (the same as the velocity of a body, initially at rest, falling distance x).

In the time interval from t to $t + \delta t$, the amount of water *lost* is $Au\delta t$, so

$$V_{t+\delta t} - V_t = -Au\delta t.$$

Divide by δt, let $\delta t \to 0$, to find

$$\frac{dV}{dt} = -Au = -A\sqrt{2gx}.$$

The volume V plainly depends on x, and the shape of the bath or sink. And since $\frac{dV}{dt} = \frac{dV}{dx}\frac{dx}{dt} = V'(x)\frac{dx}{dt}$, we can rewrite the relation as

$$\frac{dx}{dt} = -A\sqrt{2gx}/V'(x),$$

which, when solved, lets us find the time it takes for x to decrease from its initial level to zero.

Vocabulary. The *order* of a differential equation is the order of the highest derivative present—(2.1) and (2.2) above are second order, the others are first order. The *degree* is the highest power to which the derivative of highest order appears. All of the equations above are 'first degree'.

Thus the 'simplest' format for a differential equation is

$$\frac{dy}{dx} = f(x, y) \tag{2.7}$$

for some function f; it is first order, first degree. You should not necessarily expect that such a differential equation, even if it only involves standard functions, is solvable with standard functions! For example, how might you set about solving $\frac{dy}{dx} = \sin(\exp(x) + 3y)$? But numerical methods could come to our rescue in such cases, and many differential equations that arise naturally fall into one of a small number of different types, each with its own method of solution. We look at some.

2.2   First-Order Equations

(i) If Eq. (2.7) can be put in the form $\frac{dy}{dx} = A(x).B(y)$, i.e. the function $f(x, y)$ is the product of some function of x alone with another involving y alone, we term it *variables separable*, and seek to solve it via

$$\int \frac{dy}{B(y)} = \int A(x)dx.$$

To complete the job, use standard techniques of integration to find both integrals, throw in a constant of integration, and use whatever other information we have— maybe the value of y when $x = 0$—to get the relevant solution.

Example 2.3 Equation (2.3) above has its variables separable: it can be turned into

$$\int \frac{dv}{g - cv} = \int dt,$$

hence $\frac{-1}{c} \log(g - cv) = t + K$, K a constant. If the stone starts from rest, so that $v = 0$ when $t = 0$, we find $v = \frac{g}{c}(1 - \exp(-ct))$.
    Since $v = \frac{dx}{dt}$, this becomes

$$\frac{dx}{dt} = \frac{g}{c}(1 - \exp(-ct)),$$

which also has its variables separable, and leads to

$$\int dx = \frac{g}{c} \int (1 - e^{-ct})dt,$$

easily integrable giving x as a function of t. See also Example 2.6 below.

(ii) There is a cunning trick available if the function f in (2.7) is *homogeneous*, i.e. if $f(tx, ty) = f(x, y)$ whenever t is non-zero. Change the variables from (x, y) to (x, v) by writing $y = vx$. Then

$$\frac{dy}{dx} = v + x\frac{dv}{dx},$$

and the right side of (2.7) becomes $f(x, vx)$; but $f(x, vx) = f(1, v)$ by the homogeneity property, so the original Eq. (2.7) becomes

$$v + x\frac{dv}{dx} = f(1, v).$$

This can be rewritten as $x\frac{dv}{dx} = f(1, v) - v = g(v)$ (say), clearly separable to

$$\int \frac{dv}{g(v)} = \int \frac{dx}{x}.$$

Remember the cunning trick, not this formula!

Example 2.4 The stream in a river c metres wide flows at the uniform rate a metres/second. A boat travelling at constant speed b metres/sec sets off to cross the river, always steering for the point directly opposite its initial point. Describe its path. For what values of a and b can it reach its goal?

Solution. Let the river banks be the parallel lines $x = 0$ (i.e. the y-axis) and $x = c$. The boat starts at $(c, 0)$, and always aims at $(0, 0)$. Suppose it is at position (x, y) as shown in Fig. 2.2.

Look separately at the components of the velocity, first across the river (i.e. x), and then upstream (i.e. y). Because the boat is aiming at the origin, we have

$$\frac{dx}{dt} = -b\cos(\theta).$$

Also, the influence of the current means that

$$\frac{dy}{dt} = -a + b\sin(\theta).$$

Divide these out to get

$$\frac{dy}{dx} = \frac{dy/dt}{dx/dt} = \frac{-a + b\sin(\theta)}{-b\cos(\theta)} = \frac{-a + b(-y/\sqrt{x^2 + y^2})}{-b(x/\sqrt{x^2 + y^2})},$$

which simplifies to

$$\frac{dy}{dx} = \frac{a\sqrt{x^2 + y^2} + by}{bx}.$$

Fig. 2.2 The boat aims at its destination

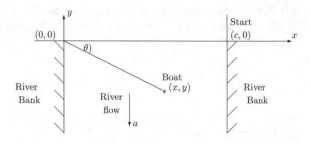

The right side is a homogeneous function, so put $y = vx$ to get

$$v + x\frac{dv}{dx} = \frac{a\sqrt{1+v^2} + bv}{b} = \frac{a}{b}\sqrt{1+v^2} + v.$$

Take $k = a/b$, the ratio of the speed of the river to the speed of the boat. Then $x\frac{dv}{dx} = k\sqrt{1+v^2}$ leading to

$$\int \frac{dv}{\sqrt{1+v^2}} = \int k\frac{dx}{x} = k\log(x) + Const.$$

To find the integral on the left side, if you have met the functions $\sinh(u)$ and $\cosh(u)$, you will be able to solve this by the substitution $v = \sinh(u)$, and some work with inverses. Otherwise, write $u = \log(v + \sqrt{1+v^2})$, and see that

$$\frac{du}{dv} = \frac{1 + (1/2)(1+v^2)^{-1/2}.2v}{v + \sqrt{1+v^2}} = \frac{1}{\sqrt{1+v^2}}.$$

Hence, since differentiation and integration are inverse processes, the integral we seek is indeed $u = \log(v + \sqrt{1+v^2})$.

Thus $\log(v + \sqrt{1+v^2}) = k\log(x) + Const$. Initially, when $x = c$, so $y = 0$, hence also $v = 0$. So $0 = k\log(c) + Const$ giving

$$\log(v + \sqrt{1+v^2}) = k\log(x) - k\log(c) = \log\left(\frac{x}{c}\right)^k.$$

Since $v = y/x$, this becomes

$$y + \sqrt{x^2 + y^2} = x^{k+1}/c^k \tag{2.8}$$

in our original notation.

This has 'solved' the equation, but it is usual to try to end up with a solution in the form $y = A(x)$ for some expression $A(x)$. To this end, rewrite (2.8) as

$$\sqrt{x^2 + y^2} = \frac{x^{k+1}}{c^k} - y,$$

and square both sides. Cancel the term y^2 on each side, leading to

$$y = \frac{1}{2}\left(\frac{x^{k+1}}{c^k} - \frac{c^k}{x^{k-1}}\right).$$

Near $x = 0$, i.e. just before completing the crossing, the second term dominates. We look at the different cases.

(i) If $k > 1$, (the river flows faster than they can row), then as $x \to 0$, so $y \to -\infty$. The boat is swept downstream.
(ii) If $k = 1$, then $y = (x^2/c - c)/2$ so, as $x \to 0$ then $y \to -c/2$. The boat lands, but half a river's width downstream of where it was aiming.
(iii) If $k < 1$, (they can row faster than the flow), then $y \to 0$ as $x \to 0$. The boat reaches its destination.

Let's check that these answers seem reasonable. If $k > 1$, when the boat is downstream of its target, and near its aim point, they are trying to row directly against the current, which is too strong.

When $k < 1$, they can indeed row faster than the opposing stream, so do reach their target.

When $k = 1$, they can row at exactly the stream speed. This case isn't important, as it will never happen than these speeds are EXACTLY equal. But even so, the answer, landing precisely $c/2$ downstream, is not easy to guess!

(iii) A *First-Order Linear Equation* is one of the form

$$\frac{dy}{dx} + P(x)y = Q(x),$$

and this too has a standard method of solution. Write $R(x) = \exp(\int P(x)dx)$, the so-called *integrating factor*: just multiply throughout by $R(x)$ leading to

$$R(x)\frac{dy}{dx} + R(x)P(x)y = R(x)Q(x).$$

This does represent progress *because* the left side is just $\frac{d}{dx}(R(x)y)$ (yes?), and so our equation becomes

$$\frac{d}{dx}(R(x)y) = R(x)Q(x).$$

Hence, integrating leads to

$$R(x)y = \int R(x)Q(x)dx.$$

Even if the function $R(x)Q(x)$ doesn't integrate to a standard form, numerical techniques can be used.

Again, do not remember this formula! Learn the method, as illustrated now.

Example 2.5 Solve

$$x\frac{dy}{dx} - 3y = x^4.$$

Solution. First rewrite it in the standard form for a first-order linear equation,

$$\frac{dy}{dx} - \frac{3y}{x} = x^3.$$

The integrating factor is $\exp(\int(-3/x)dx)) = \exp(-3\log(x)) = x^{-3}$, so multiply through by this to get

$$\frac{dy}{dx}x^{-3} - 3yx^{-4} = 1,$$

i.e.

$$\frac{d}{dx}\left(yx^{-3}\right) = 1.$$

Integration gives $y/x^3 = x + c$, i.e. $y = x^4 + cx^3$, c an arbitrary constant.

(Check: given this y, differentiate to see that $\frac{dy}{dx} = 4x^3 + 3cx^2$, so that $x\frac{dy}{dx} - 3y = 4x^4 + 3cx^3 - 3(x^4 + cx^3) = x^4$, as we wanted.)

2.3 Second-Order Equations with Constant Coefficients

There is a standard approach to solving

$$a\frac{d^2y}{dx^2} + b\frac{dy}{dx} + cy = f(x), \tag{2.9}$$

when a, b and c are constants and $a \neq 0$. First, find the general solution to the corresponding *homogeneous equation*,

$$a\frac{d^2y}{dx^2} + b\frac{dy}{dx} + cy = 0 \tag{HE}$$

by *assuming* a solution of the form $y = \exp(mx)$ for some unknown m. Then plainly (HE) leads to

$$\exp(mx)(am^2 + bm + c) = 0.$$

But, since $\exp(mx)$ is never zero, the only solutions come from solving the quadratic $am^2 + bm + c = 0$ (known as the *auxiliary equation*) in the standard way. There are three possibilities:

(a) two different real roots, m_1 and m_2;

(b) one repeated real root, m;
(c) two complex conjugate roots $u + iv$ and $u - iv$, u and v real.

All of these are dealt with in a similar way. Let A and B be arbitrary constants. In case (a), the general solution of (HE) is

$$Y_1(x) = A \exp(m_1 x) + B \exp(m_2 x);$$

in case (b), it is $Y_1 = (A + Bx) \exp(mx)$, while case in (c) it can be written as $Y_1 = \exp(ux)(A \cos(vx) + B \sin(vx))$.

The next step is to use your ingenuity to find, somehow or other, SOME solution of the original Eq. (2.9); you should pay close attention to the form of $f(x)$, and use trial and error. Let $Y_2(x)$ be some such solution. Then the general solution to (2.9) is

$$y = Y_1(x) + Y_2(x).$$

In case (a), this is $y = A \exp(m_1 x) + B \exp(m_2 x) + Y_2(x)$.

Finally, use other information, such as the values of y and $\frac{dy}{dx}$ when $x = 0$, or the values of y when $x = a$ and when $x = b$, to find the relevant values of A and B. Be clear that this last step is made only AFTER both Y_1 and Y_2 have been found!

Example 2.6 Equation (2.2) above, i.e.

$$\frac{d^2 y}{dt^2} = g - c\frac{dy}{dt},$$

comes from Newton's Law. Here y is the distance travelled in time t when a stone moves vertically, subject to air resistance. Take $c > 0$, and assume that $y = 0$ and $\frac{dy}{dt} = v_0$ when $t = 0$.

Solution. Rewrite it as

$$\frac{d^2 y}{dt^2} + c\frac{dy}{dt} = g,$$

so that the homogeneous version is

$$\frac{d^2 y}{dt^2} + c\frac{dy}{dt} = 0.$$

This leads, using $y = \exp(mt)$, to the auxiliary equation $m^2 + cm = 0$, with two different solutions $m = 0$ and $m = -c$.

Thus the general solution of the homogeneous equation is

$$y = A + B \exp(-ct).$$

For a particular solution to

$$\frac{d^2 y}{dt^2} + c\frac{dy}{dt} = g,$$

guess $y = Kt$, giving $cK = g$, i.e. $y = gt/c$. The general solution of the equation is the sum of these solutions, namely

$$y = A + B\exp(-ct) + gt/c.$$

Recall that $y = 0$ and $\frac{dy}{dt} = v_0$, when $t = 0$. Thus find A and B from $0 = A + B$ and $v_0 = -Bc + g/c$ leading to the solution

$$y = \frac{v_0}{c}(1 - \exp(-ct)) + \frac{g}{c^2}(\exp(-ct) + ct - 1).$$

This gives another excuse to use L'Hôpital's Rule. See what happens as $c \to 0$. The first term on the right side is

$$\frac{v_0(1 - \exp(-ct))}{c}.$$

Differentiate numerator and denominator (as functions of c) to get $v_0 t \exp(-ct)$ and unity, whose ratio converges to $v_0 t$. The second term is

$$\frac{g(\exp(-ct) + ct - 1)}{c^2},$$

and it will turn out that we need to differentiate numerator and denominator *twice*. Their respective second derivatives are $gt^2 \exp(-ct)$ and 2, so the second term converges to $gt^2/2$. The whole expression converges to $tv_0 + gt^2/2$.

This suggests that $y = tv_0 + gt^2/2$ solves the corresponding problem when $c = 0$, i.e *without* air resistance. Exercise 2.14 asks you to check this, using the general solution to (HE) in the case (b) of equal roots of the quadratic.

2.4   Linked Systems

Sometimes, we have several interacting quantities, all also changing with time. We look first at the *Lotka-Volterra* model of a predator–prey system.

Rabbits on an island feed on an unlimited supply of clover, and foxes feed on rabbits. So let x be the number of rabbits at time t and let y be the number of foxes. (As we wish to differentiate, we shall treat x and y as continuous variables, even though they will be integers.) In the absence of foxes, rabbits breed happily, and we expect

$$\frac{dx}{dt} = ax$$

for some $a > 0$. With foxes, the rate at which foxes and rabbits meet is proportional to the *product* xy of their numbers, so, for some $b > 0$, we have

$$\frac{dx}{dt} = ax - bxy. \tag{2.10}$$

If there were no rabbits, the foxes would just die out leading to

$$\frac{dy}{dt} = -cy$$

with $c > 0$. But eating rabbits enables breeding; and since the rate of meeting is proportional to xy, we see that

$$\frac{dy}{dt} = -cy + hxy \tag{2.11}$$

with $h > 0$. Setting the right sides of (2.10) and (2.11) to zero, i.e.

$$x(a - by) = 0 \quad \text{and} \quad y(hx - c) = 0$$

gives an *equilibrium*, as the rate of change of the size of both populations is then zero. With both species present, the only solution is $x = c/h$ and $y = a/b$.

Vito Volterra became interested in this sort of problem when his attention was drawn to the data on the proportion of predatory fish among all fish caught at the Italian port of Fiume. It rose from around 12% just before the First World War in 1914 to around 30% at the end of the war in 1918/9, before returning to its previous level. In the above model, the dramatic reduction in general fishing during the war changed the values of the parameters in Eqs. (2.10) and (2.11): the natural growth rate of prey, a, increased, the natural death rate of the predators, c, decreased, but both b and h, which apply to the interactions between the prey and predators, were not affected. Thus at the model's equilibrium point $(c/h, a/b)$, with little fishing, the first value dropped, the second increased—exactly what the observed data confirmed!

In 1910, Alfred Lotka had already developed the same system of equations, but in connection with the analysis of certain chemical reactions.

What happens if we are away from the equilibrium point? Dividing out (2.10) and (2.11), we get

$$\frac{dy}{dx} = \frac{y(hx - c)}{x(a - by)}$$

which has separable variables, hence

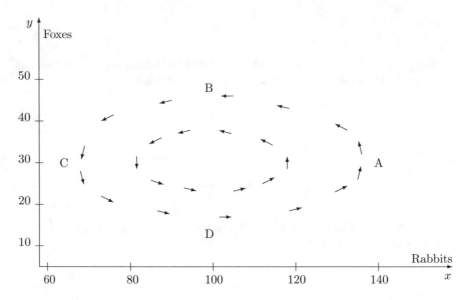

Fig. 2.3 How numbers change away from equilibrium

$$\int \frac{(a - by)dy}{y} = \int \frac{(hx - c)dx}{x}.$$

Integrate to find

$$a \log(y) - by = hx - c \log(x) + Const.$$

Take exponentials of both sides, rewrite as

$$y^a \exp(-by) = K \exp(hx)/x^c,$$

where K is a constant that can be found from the initial values of x and y.

This 'solution' is a collection of closed curves around the equilibrium point—to a first approximation, they are distorted ellipses, as you will confirm when you solve Exercise 2.15. Figure 2.3 illustrates how the two populations change when disturbed from equilibrium—for this illustration, we have chosen $a = 30$, $b = 1$, $c = 50$ and $h = 0.5$, for which $(100, 30)$ is the equilibrium. The arrows indicate the model's predictions for initial states either $(120,30)$ or $(140,30)$; in the former, the rabbit population ranges from 83 to 120, the fox population from 23 to 38, while in the latter, rabbit numbers run from 70 to 140, fox numbers from 18 to 46.

It is plain that the picture accords with our intuition. Starting with the equilibrium number of foxes, but excess rabbits (near point A, say), foxes are favoured: their numbers increase, rabbit numbers decrease, until there are too few rabbits to support the large excess of foxes (after B); both populations now decline, until rabbits are so scarce that foxes die off quickly (at C), enabling the rabbit population to recover (along CD). With sufficiently many rabbits around (after D), fox numbers increase again, and the whole cycle repeats from A.

Another linked system is the *Kermack–McKendrick* model for epidemics.

Imagine an isolated community of size n into which one person infected with an infectious disease is introduced. At time t, there are x Susceptibles (people who may catch the disease), y Infecteds (people with the disease, and spreading it) and z Removed cases (they have had the disease, and have recovered, or died, or been placed in isolation: they are no longer susceptible). Take $x + y + z = n + 1$, so that any two of x, y, z determine the third; and again treat then as *continuous* variables (even though they are not) so as to be able to differentiate them. Initially, $x = n$, $y = 1$ and $z = 0$.

When an Infected comes into contact with a Susceptible, the Susceptible *might* become an Infected (and so infectious), while as the disease runs its course, Infecteds become Removed. How might the disease spread, how many will escape infection?

The quantity xy is the number of possible encounters between a Susceptible and an Infected; the actual number of encounters in a given time period will depend on whether the population is tightly packed, or sparsely scattered, and whether the encounter leads to infection will depend on the infectiousness of the disease. For some constant $\beta > 0$, the quantity βxy describes the rate at which a Susceptible turns into an Infected; so we have

$$\frac{dx}{dt} = -\beta xy.$$

The right side also describes the rate at which the number of Infecteds increases; but Infecteds turn into Removed cases (by recovering, dying, being isolated, etc.) at a rate simply proportional to their number, so putting these two factors together we have

$$\frac{dy}{dt} = \beta xy - \gamma y$$

for some constant $\gamma > 0$. Finally, this last term also gives the rate at which Removed cases increase, so

$$\frac{dz}{dt} = \gamma y.$$

From the first two equations, divide out to find

$$\frac{dy}{dx} = \frac{\beta xy - \gamma y}{-\beta xy} = \frac{\rho - x}{x} \tag{2.12}$$

where $\rho = \gamma/\beta$. This has separable variables, so we integrate to find

$$\int dy = \int \frac{\rho - x}{x} dx,$$

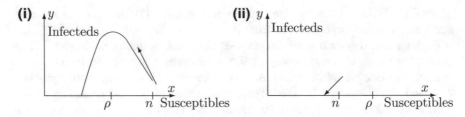

Fig. 2.4 How numbers of infecteds and susceptibles change

which gives $y = Const - x + \rho \log(x)$. Since $y = 1$ when $x = n$, the general rela-
tionship is $y = n + 1 - x + \rho \log(x/n)$. The epidemic is over when $y = 0$, so we
deduce that the number who escaped infection is x_1, found from the equation
$x_1 = n + 1 + \rho \log(x_1/n)$.

Suppose first that $n > \rho$. Initially $x = n$, so (2.12) shows that $\frac{dy}{dx}$ is initially negative; and since x can only decrease, this negative slope means that y, the number
infected, begins to increase, continues to do so while $x > \rho$, but then decreases until
$y = 0$ and the epidemic ends, as in Fig. 2.4(i). The formula for x_1 above shows that,
to a first approximation, if there are $\rho + \delta$ Susceptibles initially, there are about $\rho - \delta$
when the epidemic is over. See Exercise 2.16.

By contrast, if $n < \rho$, (2.12) shows that the initial value of $\frac{dy}{dx}$ is positive, so y,
already tiny, tends to decrease as x decreases, and the disease swiftly dies out, with
no epidemic, as in Fig. 2.4(ii).

The crucial quantity $\rho = \gamma/\beta$ is termed the *threshold* for this system. If $n < \rho$, the
disease is expected to die out quickly, whereas if $n > \rho$, an epidemic will occur. (Of
course, this analysis is based on *averages* so, in practice, the size of any epidemic will
be some random number, whose average we have found. In case (ii), this average
is tiny, so there is no real chance of a large outbreak; in case (i), random chance
MIGHT eliminate the disease before it has time to take hold, but if it does not die
out quickly, there is a good possibility of a substantial epidemic.)

So overall, to prevent epidemics, we wish to ensure that $n < \rho$. One obvious
way is vaccination, to reduce the value of n. Other ways are to increase $\rho = \gamma/\beta$:
either by increasing γ—e.g. remove Infecteds quickly, or help them recover quickly
(or cull them if this is a foot-and-mouth epidemic in cattle!); *or* by reducing β—
perhaps disperse the population, or close schools, or postpone sports events or pop
concerts to reduce the frequency of contact between Infecteds and Susceptibles. Even
vaccination that is partly effective by reducing the infectiousness of a disease will
also reduce β. Our mathematical model points to ways of reducing the frequency
and severity of epidemics, and enables us to forecast how cost-effective possible
measures to combat the disease might be.

2.5 Exercises

(Several of these Exercises, and the chapter examples, are adapted from material in the excellent book by George Simmons, that also contains interesting historical information on this subject and its pioneers.)

2.1 A model for the growth of bacteria subject to competition is

$$\frac{dx}{dt} = Kx(1 - \frac{x}{N}).$$

Here t represents time and x is the amount of bacteria. Suppose that, at time $t = 0$, then $x = N/4$, while at $t = 1$, then $x = N/2$. At what time will we find $x = 3N/4$?

2.2 The principle of radiocarbon dating is that air-breathing plants absorb carbon dioxide only while they are alive: radiocarbon is a radioactive isotope of carbon, and its proportion in the atmosphere has long been in equilibrium. It decays after the plant's death, so that comparing the proportion of radiocarbon in a piece of old wood with the current proportion in the atmosphere gives an estimate of the length of time the tree has been dead.

Suppose matter decays at a rate proportional to its quantity. Write down the form of the differential equation for $x(t)$, the amount of matter at time t, and deduce that $x(t) = x(0) \exp(-kt)$ for some constant $k > 0$.

It takes around 5600 years for half the initial amount to decay (hence the term *half-life*). Estimate the age of a wooden object whose proportion of radiocarbon is (i) 75% (ii) 10% of the current amount. (Round your answers sensibly.)

2.3 Snow has been falling steadily for some time when the snowplough begins to clear the road at noon. It removes snow at a constant rate, and has covered two miles by 1-00 pm, two further miles by 3-00 pm.

We have argued that, if snow began T hours before noon and fell at rate s/unit time, while the plough cleared V/unit time and had moved distance x in time t, then

$$s(T + t)\frac{dx}{dt} = V.$$

Solve this equation with the given conditions, and deduce the time that snow started to fall.

2.4 *Newton's Law of Cooling* states that the rate an object cools is proportional to the temperature difference between it and its surroundings. A rock is heated to $120\,°C$, and placed in a large room kept at a constant temperature at $20\,°C$. The temperature of the rock falls to $60\,°C$ after an hour; how much longer does it take to cool to $30\,°C$?

2.5 A spherical mothball evaporates uniformly at a rate proportional to its surface area. Hence deduce a differential equation that links its radius with time. Given that the radius halves from its initial value in one month, how long will the mothball last?

2.6 We have shown that if the water level in a sink or bath is $x > 0$, then

$$\frac{dx}{dt} = -\frac{A\sqrt{2gx}}{V'(x)},$$

where V is the volume of water, V' is its derivative, t is time and A is the cross-sectional area of the plughole.

(i) Suppose a bath can be taken as having a rectangular horizontal cross section of fixed area B. How long will it take to empty from depth $D > 0$? (Your answer should just involve D, B, A and g.) Now suppose a blockage *halves* the cross-sectional area of the plughole, and you decide to have a really luxurious bath by doubling the amount of water: what difference does all that make to the time to empty the bath?

(ii) Take the plughole to be a circle, radius $r > 0$ and suppose a sink is a perfect hemisphere of radius $R > 0$. Show that the volume of water when the depth is $x > 0$ is $V = \pi x^2 (3R - x)/3$. Deduce that the time for the bowl to empty from full is $\dfrac{14R^{5/2}}{15r^2\sqrt{2g}}$.

(iii) Let f be a smooth and strictly increasing function with $f(0) = 0$. A sink has the shape of the graph $y = f(x)$, for $0 \le x \le c$, being rotated around the y-axis. Give the form of f that corresponds to the water level falling at a *constant* rate.

(A *clepsydra* is a water clock consisting of a sink with a small hole to allow water to escape. It was used in ancient Greek and Roman courts to time the speeches of lawyers. It is plainly an advantage in such an implement to have the water level falling at a uniform rate.)

2.7 Find y as a function of x when $y = 1$ when $x = 1$, and

$$\frac{dy}{dx} = \frac{3xy + 2y^2}{x^2}.$$

2.8 Find y as a function of x for $x > 0$ when

$$\frac{dy}{dx} - \frac{5y}{x} = x,$$

and $y = 1$ when $x = 1$.

2.9 Suppose $\frac{dy}{dx} + P(x)y = Q(x)y^n$ for some n, with $n \ne 0$ and $n \ne 1$ (known as *Bernoulli's Equation*). Define $z = y^{1-n}$, and show that the resulting differential equation linking z and x is linear.

Hence solve $x\frac{dy}{dx} + y = x^4 y^3$, given that $y = 1$ when $x = 1$.

2.10 In the model of rowing across a river described in this chapter, suppose that we can row twice as fast as the river flows. Use the result

$$y = \frac{1}{2} \left(\frac{x^{k+1}}{c^k} - \frac{c^k}{x^{k-1}} \right)$$

and the relation

$$\frac{dx}{dt} = \frac{-bx}{\sqrt{x^2 + y^2}}$$

to find an equation linking x and t. Deduce how long it takes to cross the river.

2.11 A 100-L tank is initially full of a mixture of 10% alcohol and 90% water. Simultaneously, a pump drains the tank at 4 L/s, while a mixture of 80% alcohol and 20% water is poured in at rate 3 L/s. Thus the tank will be empty after 100 s (yes?). Assume that the two liquids mix thoroughly, and let y litres be the amount of alcohol in the tank after t seconds; explain why the equation

$$\frac{dy}{dt} = 2.4 - \frac{4y}{100 - t}$$

holds for $0 \le t < 100$. Find y as a function of t; hence deduce that the maximum amount of alcohol in the tank occurs after about 34 s, and is is about 39.5 L.

2.12 Twenty-five grams (gm) of salt are dissolved in fifty litres of water, and poured into tank A. From time $t = 0$, pure water is poured into this tank at the rate of two litres/second, and simultaneously two litres/second drain from the tank. Thus the tank always contains 50 L of liquid, thoroughly mixed. Let x (gm) denote the amount of salt in tank A at time $t \ge 0$; justify the equation

$$\frac{dx}{dt} = -x/25.$$

The liquid that drains from tank A pours into tank B, which initially has 50 L of pure water, and two litres/second drain from tank B. Thus tank B also always contains 50 L of liquid, thoroughly mixed. Let y (gm) denote the amount of salt in tank B at time $t \ge 0$. Explain briefly why

$$\frac{dy}{dt} = x/25 - y/25.$$

Hence find x, then y. At what time is the amount of salt in tank B at its maximum?

2.13 A chain of length $L = 150$ cm has constant density ρ/gm/cm. It rests on a smooth horizontal table with 30 cm hanging over the edge, and is released. Neglect friction. It slides off the table according to Newton's Law:

$$\text{Mass} \times \text{Acceleration} = \text{Force.} \tag{2.13}$$

The Mass is constant at $L\rho$ gm, but the Force varies according to the amount of overhang; for an overhang of length x, the Force will be $x\rho g$. Set up the relevant differential equation, state the initial conditions and find how long it takes for the chain to fall off the table. (Take $g = 10$ m/s/s.)

2.14 Solve the equation

$$\frac{d^2y}{dt^2} = g,$$

where y represents the distance travelled at time t when a stone is thrown vertically, with initial velocity v_0, in the absence of air resistance.

2.15 Using the notation in this chapter for the Lotka–Volterra model, write $x = c/h + X$ and $y = a/b + Y$, where X and Y are assumed *small* (i.e. (x, y) is close to the equilibrium point $(c/h, a/b)$). Find the exact equations for X and Y that correspond to (2.10) and (2.11), and then, neglecting the terms involving the product XY, divide out these equations to obtain dY/dX (approximately). Show that the exact solutions to this approximate equation form a family of ellipses centered at the equilibrium point; select a representative member of this family and show (by marking arrows at a selection of points) how the population sizes are expected to change if the two populations are disturbed from the equilibrium point.

2.16 Suppose one infected individual enters a population of size 1100, when the relevant value of the threshold is $\rho = 1000$. According to the Kermack–McKendrick model, how many people will be infected before the epidemic dies out?

2.17 In a model of warfare, let the sizes of the opposing armies at time t be $x(t)$ and $y(t)$, and let a, b be positive constants. Then the equations

$$\frac{dx}{dt} = -by, \qquad\qquad \frac{dy}{dt} = -ax$$

might describe how the conflict develops; for example, b will be larger when the y−army has strong offensive capabilities and the x−army has weak defences, and vice versa for the interpretation of a.

Divide one equation by the other to eliminate t, and deduce *Lanchester's Square Law*, i.e. $a(x(0)^2 - x(t)^2) = b(y(0)^2 - y(t)^2)$.

Assume that fighting continues until one side is annihilated: show that the x-army wins whenever $ax(0)^2 > by(0)^2$, and that if this holds, the final size of the x-army is $\sqrt{x(0)^2 - (b/a)y(0)^2}$.

Now suppose the armies have equal initial sizes, and the same offensive and defensive capabilities so that $a = b$, but the x-army has split the other into two equal parts, and will fight them sequentially. What is the size of the x-army after the first battle? After the second?

Give a similar analysis for when the y−army has been split into three equal parts, tackled sequentially. Are your two answers consistent with the commonly stated principle of offensive warfare that 'The more you use, the less you lose.'?

2.18 As a parallel to the Kermack–McKendrick model for the spread of epidemics, Daryl Daley and David Kendall proposed a model for the spread of rumours. A homogeneously mixing community, initially contains one person (a Spreader) who knows a rumour, and N who do not (the Ignorants). When a Spreader meets an Ignorant, the Ignorant learns the rumour and becomes a Spreader; but if a Spreader attempts to tell the rumour to another person who already knows it, both of them believe the rumour to be 'old hat', and decide to cease spreading it: people who know the rumour but no longer spread it are Stiflers.

At time t, let x be the number of Ignorants, y the number of Spreaders, and z the number of Stiflers; thus $x + y + z = N + 1$. By considering the outcomes of encounters between the different types of person, justify the equations

$$\frac{dx}{dt} = -Axy,$$

$$\frac{dy}{dt} = Axy - Ayz - Ay(y - 1),$$

where $A > 0$ is a constant. Deduce an expression for $\frac{dy}{dx}$, and hence find y in terms of x. Show that, when the rumour dies out, the number of Ignorants, X, satisfies the equation

$$0 = 1 + N \log(X/N) + 2(N - X).$$

Hence show that, when N is large, this model predicts that about 80% of the Ignorants will learn the rumour, and, in contrast to the epidemic model, there is no 'threshold theorem'.

References and Further Reading

Daley D J and Kendall D G (1965) Stochastic Rumours. *Journal of the Institute of Mathematics and its Applications* 1 pages 42–55

Kermack W O and McKendrick A G (1927) A contribution to the Mathematical theory of Epidemics. *Proceedings of the Royal Society A* 115 pages 700–721

Lanchester F W (1916) Aircraft in warfare: the Dawn of the Fourth Arm. Constable and Co.

Lotka A J (1910) Contribution to the Theory of Periodic Reactions. *Journal of Physical Chemistry* 14 pages 271–4

Simmons G F (1972) Differential Equations, with Applications and Historical Notes. McGraw-Hill.

Volterra V (1926) Fluctuations in the abundance of species considered mathematically. *Nature* 118 pages 558–560

Chapter 3
Sport and Games

3.1 Lawn Tennis

In this popular pastime, whether played as singles (i.e. one against one) or doubles (one pair against another), the right to serve is a substantial advantage among good players. A *match* consists of a sequence of *sets*, each set consists of a number of *games*, and within a game, the same player serves to every *point*. A game ends when one side has won at least four points, and has also scored at least two points more than their opponents. Suppose the serving side tends to win a proportion p of the points; what proportion of service games should they expect to win?

Instead of counting the points in the usual sequence 0, 1, 2, 3, etc., tennis replaces 1 by 15, 2 by 30 and 3 by 40 (!), so we will use this convention. For the simplest plausible model, we assume that the outcome of any point has no effect on the outcomes of other points, the serving side winning a point with a fixed probability p, so that the receiving side wins any point with probability $q = 1 - p$. We answer our question by listing the distinct ways in which the serving side can win the game, finding their respective chances, and summing these values.

Plainly, if they have won the game, the serving side did win the final point. The distinct scores before this final point are (leader first):

(a) 40–0, when winning the first 3 points, probability p^3;
(b) 40–15, when winning any 3 of the first 4 points, with possible orders WWWL, WWLW, WLWW, LWWW, each with probability p^3q. The total chance is $4p^3q$.
(c) 40–30 when winning any 3 of the first 5 points. There are 10 possible orders, each with chance p^3q^2, giving a total of $10p^3q^2$.
 These are the chances of *reaching* each score: multiply by p and sum, to get the chance of winning the game from any of those scores.
(d) Finally, the score reaches 'Deuce', or 40–40, if each side wins 3 of the first 6 points. There are 20 possible orders (e.g. WWLWLL), each with probability p^3q^3, so the chance of *reaching* Deuce is $20p^3q^3$. After that score, a lead of two points automatically ensures that at least four points have been scored, so take x as the chance the server wins when the current score is Deuce. To be the winner,

© Springer Nature Switzerland AG 2019
J. Haigh, *Mathematics in Everyday Life*,
https://doi.org/10.1007/978-3-030-33087-3_3

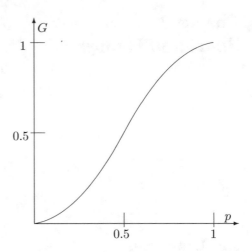

Fig. 3.1 From a point to a game

either the server wins the next two points (probability p^2), *or* the next two points are shared (probability $2pq$), bringing the score back to Deuce, from where the server goes on to win. Overall, we have

$$x = p^2 + 2pqx,$$

leading to $x = p^2/(1 - 2pq)$.

Putting them all together, the chance the server wins the game is (Fig. 3.1)

$$G = (p^3 + 4qp^3 + 10q^2p^3)p + 20p^3q^3\frac{p^2}{1 - 2pq} = p^4(1 + 4q + 10q^2) + \frac{20p^5q^2}{1 - 2pq}.$$

If $p = 1/2$, you can check that this leads to $G = 1/2$, as we would expect, by symmetry. Suppose $p \neq 1/2$. Put the whole expression for G over the denominator $(1 - 2pq)$, and note that since $p + q = 1$, then $1 = (p + q)^2 = p^2 + 2pq + q^2$, so $1 - 2pq = p^2 + q^2$. Hence

$$G = p^4\frac{(1 + 4q + 10q^2)(1 - 2pq) + 20pq^3}{p^2 + q^2} = p^4\frac{1 + 4q + 10q^2 - 2pq - 8pq^2}{p^2 + q^2}.$$

The numerator can be written $p^4(1 + 2q + 4q^2 + 8q^3) = p^4(1 - 16q^4)/(1 - 2q)$. Also $1 - 2q = (p + q) - 2q = p - q = (p - q)(p + q) = p^2 - q^2$, which leads to the elegant answer

$$G = \frac{p^4 - 16p^4q^4}{p^4 - q^4}.$$

If $p = 1/2$, both numerator and denominator are zero, but L'Hôpital's Rule shows that, as $p \to 1/2$, so also $G \to 1/2$ (which makes sense). The Exercises explore some consequences of this formula.

Service Tactics. Players usually have two types of serve, call them Fast (F) and Steady (S). With F, the chance of a fault is higher, but if it is not a fault, it is more likely to win the point. So players, having two serves, have a choice of four possible tactics: FF, FS, SF, SS (where e.g. FS means Fast serve first then, if it is a fault, use a Steady serve). You often see FS, sometimes FF, even SS, but (almost) never SF. Is there a good reason for this?

Let f and s be the respective chances a Fast or Steady serve wins the point, if it is not a fault; we have $f > s$. And let x, y be the chances the Fast and Steady serves are good; we expect that $x < y$.

Consider the intention to use FF: the server wins the point *either* when the first serve is good, and leads to winning the point (chance xf) *or* when the first serve is a fault, the second serve is good, and the point is subsequently won (chance $(1-x)xf$). Overall, the winning chance is

$$A = xf + (1-x)xf = xf(2-x).$$

Similar arguments lead to winning chances of $B = xf + (1-x)ys$ with FS, $C = ys + (1-y)xf$ with SF and finally $D = ys + (1-y)ys = ys(2-y)$ with SS. But

$$B - C = xf + ys - xys - (ys + xf - xyf) = xy(f - s) > 0$$

because $f > s$. Using FS *always* gives a higher winning chance than SF, and so SF should indeed never be used.

A server should logically choose those tactics that make the chance of winning any point as large as possible. Write $R = (xf)/(ys)$, the ratio of the chances of winning the point with one Fast Serve to one Steady serve, and suppose that $x < y$. Exercise 3.4 asks you to show that FF gives the best chance if $R > 1$, that FS is best if $1 > R > 1 + x - y$, and SS is best if $R < 1 + x - y$.

Example 3.1 When Anna uses her Fast serve, she has a 70% chance of winning the point, if that serve is good. However, her Fast serve is unreliable—it is a fault 60% of the time. If, without changing the potency of the serve, she could make it a valid serve half the time, what would be the change in her success rate when she decided to use her Fast serve on both service opportunities (if necessary)? Or would she rather have the same reliability, but improve her Fast serve so that it won 80% of the time, if good?

Solution. She plans to use FF, with chance $A = xf(2-x)$ of winning the point; currently $f = 0.7$, $x = 0.4$ so $A = 0.448$. If x increases to 0.5, A increases to 0.525. But if f increases to 0.8, while x remains at 0.4, then $A = 0.512$; given the choices, it is preferable to increase the reliability of the Fast serve.

Match Structure The winner of a match, or even a set, might well score fewer points than her opponent. For, suppose Anna wins a set by six games to four: in each game she won, the points score was 4–2, but whenever she lost it was 0–4: so Anna won 24 points in total, but her opponent won 28 points.

If the prime purpose of a tennis match was to identify which of two players is superior, a quite different format would be recommended. Some suitable points total, say 216, would be chosen: each player would have 108 serves, 54 from each end, alternating in blocks of six to give changeovers and rests, the winner being the one who scored most points (with some sensible rule in the case of equal scores). Such matches would often be boring—one player establishes a comfortable lead, and coasts to victory.

But the structure of a tennis match prevents such complacency, and gives much more enjoyment to spectators. 'Crucial' points occur frequently—the next point may decide who wins a game, the next game who wins a set, the next set may decide the whole match. The general dominance of the serve means that particular attention is paid to scores such as 15–40, where the receiver is said to hold two *break points*, i.e. should she win either of the next two points, she will win the game (against the odds). Match analyses often focus on how many breakpoints each player had, and how often they were then successful.

A strong server, leading 40–0, feels very comfortable, with little concern if she loses the next point. Some points are more important in determining the match-winner than others: we might measure the *importance* of a point as the difference winning or losing it makes to the outcome of a game. Pancho Gonzales, probably the best male player in the 1950s, claimed that the most important point when serving was if he found himself 15–30 down: he argued that losing the next point would be very serious—he would be 15–40 down—but winning would take him level, and his powerful serve would clinch the game.

Morris (1977) pointed out that his logic was faulty. For, compare being 15–30 down to being 30–40 down: winning the next point would take the score to 30–30 or 40–40 respectively—but these are equivalent as, in each case, the game winner will be the next player to take a two-point lead. Losing the point from 15–30 is indeed serious—but losing the next point from 30–40 is catastrophic, as the game is lost! So the *difference* between winning and losing the next point must be greater at 30–40 than at 15–30. 30–40 is always more important.

It is possible to extend this analysis by seeking to measure the 'importance' of the next point at any score, and hence try to identify the most important point in a game. However, for most scores, the answer will depend on p, the chance the server wins any particular point, whereas the analysis just given is quite independent of that notion.

Many other sports also have a structure that encourages crucial points. In table tennis, the current rule is that to win a *set*, you must score at least 11 points, with a two-point lead, and the match may be best of five, seven or nine sets: previously, sets were won by the first player to score 21 points, with a two-point lead, and matches were best of three (or five) sets.

3.2 Rugby

In rugby (League or Union), a team scores a 'try' by grounding the ball over the opposition's goal line AB, at the point X, say, in Fig. 3.2. They then attempt a 'conversion', i.e. to kick the ball between the posts C and D (and over the crossbar). This kick is taken from *any* point in the field of play on the line XE, perpendicular to the goal line. Assume that the kicker can always clear the crossbar, and when X is wide of the posts, seeks to select Y on XE to maximise the angle θ =CYD. What is the optimal place for Y?

Given Y on XE, there is a unique circle through C, D and Y. The centre of this circle, Z, is plainly on the perpendicular bisector of the goal line CD. Elementary geometry tells us that the angle CZD is just 2θ, so our task is equivalent to making CZD as large as possible. This happens when Z is as close to CD as possible, i.e. the radius ZD is as small as possible. All radii have equal length, so this is the same as making ZY as small as possible: but Z and Y are on parallel lines, so we need ZYX to be a right angle, which happens when XE is a tangent to the circle.

This gives a recipe: choose that point Y on XE such that the line XE is a tangent to the circle through C, D and Y. Exercise 3.8 invites you to practice your skill at co-ordinate geometry by finding a suitable equation to describe how Y varies with X.

Example 3.2 Suppose the posts of a rugby field are 5.5 m apart, and a try is scored 10 m wide of the nearer post. To maximise the angle between the conversion point and the two posts, how far from the goal line should the kick be taken?

Solution. Use the recipe above: take the origin as the centre of the goal line, and the $x-$axis as the goal line. The ball is grounded at the position (12.75, 0), and the centre of the circle is at (0, z), say. Then the length of the radius will be 12.75, and also will be $\sqrt{z^2 + 2.75^2}$ (Pythagoras). Hence

$$12.75^2 = z^2 + 2.75^2,$$

meaning that $z = \sqrt{155}$, i.e. take the kick 12.45 m from the goal line.

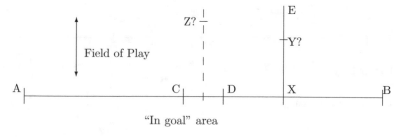

Fig. 3.2 From where should we take the kick?

3.3   The Snooker Family

In snooker (or billiards), the *cue ball* is aimed at a particular place on the *object ball* with the intention that the object ball be sent off at an angle ϕ from the straight line joining the centres of the two balls. The radius of each ball is $r > 0$, the distance between the centres of the two balls is $d > 2r$: we seek a formula for the angle θ at which the cue ball should be sent to achieve this goal. (Assume θ is small enough for the cue ball to actually strike the object ball.)

Take the centre of the cue ball as the origin, and the line joining the balls as the x-axis, so that the centre of the object ball is at $(d, 0)$, as in Fig. 3.3. Then the co-ordinates of the point of impact, P, are $(d - r\cos(\phi), r\sin(\phi))$, and so the centre of the cue ball at the time of impact is at $(d - 2r\cos(\phi), 2r\sin(\phi))$. Hence

$$\tan(\theta) = 2r\sin(\phi)/(d - 2r\cos(\phi))$$

is the EXACT relation between θ and ϕ that we seek. Exercise 3.9 asks you to use this formula to see how errors in the direction you send the cue ball translate into errors in the direction of the object ball.

Suppose a carom billiards table (i.e. without pockets) has the shape of an *ellipse*, whose standard equation is

$$\frac{x^2}{a^2} + \frac{y^2}{b^2} = 1,$$

with $a > b > 0$. Its *eccentricity* is the value $e = \sqrt{(1 - b^2/a^2)}$, and the points $(\pm ae, 0)$ are called its *foci*. Exercise 3.10 asks you to look at what happens when a cue ball is placed on one focus, and struck to hit the cushion at an arbitrary point $P = (x_1, y_1)$. To aid you, note that if we differentiate the above equation, we get

$$\frac{2x}{a^2} + \frac{2y}{b^2}\frac{dy}{dx} = 0,$$

Thus the slope of the *tangent* at P is $-(x_1/a^2)/(y_1/b^2) = -x_1 b^2/(y_1 a^2)$.

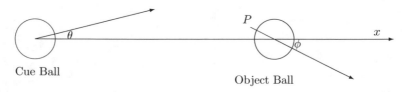

Cue Ball   Object Ball

Fig. 3.3   Aiming the cue ball at the object ball

3.4 Athletics

Throwing events. A shot-putter releases the shot at V metres/second, from height h metres, and at an angle θ above the horizontal. How far will it go before it hits the ground, assuming that air resistance can be neglected?

At time $t \geq 0$, let x be the horizontal distance travelled, and y the vertical distance above the ground, and take the acceleration due to gravity as g metres/sec/sec. When $t = 0$, we have $x = 0$, $\frac{dx}{dt} = V \cos(\theta)$, $y = h$ and $\frac{dy}{dt} = V \sin(\theta)$. Since we neglect air resistance, there is no horizontal force acting on x, so (Newton)

$$\frac{d^2 x}{dt^2} = 0.$$

The general solution is $x = A + Bt$, A and B being constants; using the initial conditions, we see that $x = Vt \cos(\theta)$.

Newton's Law also tells us that

$$\frac{d^2 y}{dt^2} = -g.$$

The corresponding homogeneous equation similarly has general solution $y = C + Dt$, and it is easy to guess a particular solution as $y = -gt^2/2$. Thus the general solution is $y = C + Dt - gt^2/2$, which leads to

$$y = h + Vt \sin(\theta) - gt^2/2$$

(so long as, of course, $y \geq 0$).

Write $V \sin(\theta) = U$. The shot hits the ground at that time $t > 0$ when $y = 0$, i.e. a solution of

$$gt^2 - 2tU - 2h = 0.$$

The positive root of this quadratic is

$$t = T = (U + \sqrt{U^2 + 2gh})/g = \frac{V \sin(\theta)}{g} \left(1 + \sqrt{1 + \frac{2gh}{V^2 \sin^2(\theta)}} \right).$$

Thus the distance thrown is the value of $x = VT \cos(\theta)$, i.e.

$$\frac{V^2 \sin(2\theta)}{2g} \left(1 + \sqrt{1 + \frac{2gh}{V^2 \sin^2(\theta)}} \right).$$

For given values of V, g, h, we can use differentiation to find the optimal angle of projection. Exercise 3.11 asks you to show that this comes from $\sin^2(\theta) = 0.5V^2/(V^2 + gh)$. However, because of the way human joints are constructed, the achievable value of V depends on the angle θ, so this result is less useful than it first appears.

But it does lead to the optimal value of θ for a given V, and we can explore the effect of small changes in the values of the parameters V, h and g. Note that the value of g varies according to latitude and altitude, so certain locations are better than others for breaking records. See Exercise 3.11

This analysis would also apply to hammer-throwing, but in both the discus and the javelin, there is considerable skill in using the aerodynamic properties of the missile to achieve long distances, so this factor should be built into any model. And in the ubiquitous way in which mathematical arguments constructed for one purpose often apply elsewhere, the analysis above could also be used to assess how far a cricket ball would travel in the air when thrown, or hit with a bat; see Exercise 3.12.

Decathlon/Heptathlon. In these multi-event competitions, the separate performances in the individual components are converted into a score, and these scores are then totalled to give a single number. For example, let X be the time in seconds taken for the men's 100 m: the corresponding score is

$$25.4347(18 - X)^{1.81},$$

rounded down to the nearest integer below. Thus a time of 10 s would score 1096 points, 11 s would score 861 points, and any time of 18 s or more would score zero points.

All track events have a score with the format $C(A - X)^B$, where C, A and B are carefully chosen constants. There is an obvious difference in format for field events, where the *greater* the distance or height, the better: a formula of the form $C(X - A)^B$ is used. For example, the score in the men's High Jump for a height of X cm is

$$0.8465(X - 75)^{1.42},$$

again rounded down to the nearest integer below: so jumping 220 cm would score 992 points, just 2 m scores 803 points, and failing to reach at least 75 cm gives a zero score.

Both formats have the feature of some 'baseline' performance A, worse than which gives a zero score; C can be adjusted so that the mean scores of competitors in each event are fairly close. Mathematically, the most interesting parameter is B. In the men's decathlon, its values for the three throwing events range for 1.05–1.1, for the three jumps the range is 1.35–1.42, while for the four track events the range is 1.81–1.92. Crucially, all these values exceed unity: that means that the increase in decathlon (or heptathlon) points for a better performance in any event is *higher* than linear; and the larger the value of the exponent B, the larger the points benefit of a given percentage improvement in performance.

To illustrate: a one percent reduction in the 100 m time from 10 to 9.9 s would gain an extra 2.3% (25 points), a one percent increase in the High Jump, from 220 to 222 cm would add 1.9% (19 points), and in the men's shot put, with formula

$51.39(X - 1.5)^{1.05}$, a one per cent increase in X from 20 to 20.20 m would add some 1.1%, just 12 more points.

The athletics authorities seek to ensure that competitors specialising in a particular single event do not have an unreasonable advantage: thus when fibreglass replaced bamboo in the pole vault, leading to a considerable increase in the heights reached, the scoring system was adjusted. Changes are also made when (as happens from time to time) the specification of the javelin is altered, for safety reasons.

Plainly there is an element of arbitrariness in the choice of the values of the parameters A, B and C; no one would claim that the actual values used are conclusively optimal.

3.5 Darts

Figure 3.4 shows a standard dartboard, designed by Brian Gamlin in 1896.

The Bull scores 50 points, the Inner (around the Bull) scores 25; in each segment, the outer ring scores double, the inner ring scores treble, so the maximum score with one dart is 60, from treble 20. In the standard game, players alternately throw three darts in succession, the winner being the first to reach a total of exactly 501, finishing with a double (including the Bull as double 25).

There are $20! \approx 2.4 \times 10^{18}$ ways could be described as $\{1, 2, \ldots, 20\}$ round the segments, or $19!/2 \approx 6.1 \times 10^{16}$ distinct ways if rotations and reflections of a given ordering are not distinguished. Is this the best possible arrangement, or can we improve on it?

Fig. 3.4 Gamlin's standard dartboard

Players seek large scores in the initial stages, so one way to punish slight errors is to have low numbers next to higher ones. Reading clockwise from the top, let the numbers be $\{x_1, x_2, \ldots, x_{20}\}$, take $x_{21} = x_1$ and write

$$S = \sum_{i=1}^{20} |x_{i+1} - x_i|,$$

the sum of the absolute values of the differences between adjacent segments. S will tend to be large when high numbers are adjacent to low numbers, and any of the $9!10!/2 \approx 6.6 \times 10^{11}$ arrangements in which the ten highest numbers alternate with the ten lowest will give $S = 200$, the maximum possible. See Exercise 3.14 (ii); Gamlin's board scores 198 on this measure. David Singmaster (1980) suggested that the *sum* of any pair of adjacent numbers be as uniform as possible as a way to punish small errors; David Percy (2012) modified this idea with the proposal that, so far as possible, the total score within any set of r adjacent numbers should be similar all around the board. Both of them found alternative arrangements that were 'optimal' under sensible criteria offered. But despite these attractive mathematical ideas, Gamlin's board faces no serious challenge.

In some games, it would be convenient if odd and even numbers alternated. But see Exercise 3.14 (iii)!

Example 3.3 Describe a way to score 501, finishing on a double, with nine throws. Explain why it is not possible to 'finish' in three darts from 159, but show how 158, 157 and 156 do have three-dart finishes. Hence explain why finishes with three or fewer darts are possible for all targets in the range 2–155.

Solution. One way is to score 60 with 7 darts, then 57, finish with double 12. To finish from 159 is not possible, *because*: (a) even with a Bull finish, the first 2 darts would need to total exactly 109, so must include an odd number; using 57 then requires an additional 52, not possible, using 51 asks for 58, not possible; and (b) with a 40 to finish, the first 2 darts would need to total 119, plainly not possible.

We can get 158 as $60 + 60 + 38$, 157 as $60 + 57 + 40$ and 156 as $60 + 60 + 36$. Since the three successive scores 156, 157, 158 are possible and use two treble beds, by changing one treble bed to the next lower number, we can go all the way from 155 to 44 (not necessarily efficiently, but who cares?). And any number from 2 to 43 is plainly possible in at most 2 darts.

In the standard game, throwing first is plainly an advantage. To assess the size of this advantage, we take two equally good players, Phil and Andy, say, and assume that whoever throws first wins that game with probability $p > 0.5$. In darts terminology, each such game is termed a *leg*, a *set* may be the best of three or five legs, and a *match* could consist of almost any (odd) number of sets. Within any set, the right to throw first to a leg alternates. If Phil throws first in a set of three legs, what is the chance he wins that set?

Write W to indicate a leg Phil wins, L to mean he loses, and write $q = 1 - p$ as usual. As with our approach in the model of lawn tennis, we assume that the outcome of any leg does not affect the outcome of other legs. Then the set outcomes where Phil wins, with their respective probabilities in parentheses, are WW (chance pq), WLW (p^3) and LWW ($q^2 p$), so his overall winning chance is their sum, $pq + p^3 + q^2 p = p(2 - 3p + 2p^2) = g(p)$, say.

Let's check this makes sense—it should evaluate to 0.5 when $p = 0.5$, by symmetry—and it does. Also, $g(0) = 0$ and $g(1) = 1$, as are plainly essential. We can construct the table

p	0.5	0.6	0.7	0.8
$g(p)$	0.5	0.552	0.616	0.704

For a set of *five* legs, the list of outcomes leading to Phil winning the set when he throws first in the first leg is: WWW, $WWLW$, $WLWW$, $LWWW$, $WWLLW$, $WLWLW$, $WLLWW$, $LWWLW$, $LWLWW$ and $LLWWW$ with respective chances $p^2 q$, pq^3, etc. Their sum can be simplified to

$$g(p) = p(3 - 9p + 16p^2 - 15p^3 + 6p^4).$$

The usual checks confirm that $g(0)$, $g(0.5)$ and $g(1)$ have the correct values. The corresponding table for 5-leg sets is

p	0.5	0.6	0.7	0.8
$g(p)$	0.5	0.539	0.585	0.654

Again, throwing first carries a significant advantage towards winning the set, but, as we would expect, a smaller advantage with the longer set.

We now seek to assess the value of p among top-class players. Such a player will average around 100 on each set of throws, so taking some 4 or 5 visits to get near to 501, and perhaps two more darts, on average, to hit the finishing double. Overall, an average of 5 or 6 visits looks plausible. A nine-dart finish (three visits) is rare enough to be effectively discounted, so to give the desired average, and using round figures, I offer the table

Visits	4	5	6	7	8
Chances	0.1	0.3	0.3	0.2	0.1

as a fair assessment of the chances of a top-class darts player reaching the target of 501 in the number of visits to the oche indicated.

Assume this table holds: if you go first, you win whenever you need n visits, and your opponent needs *at least* n visits. Break this down:

(a) $n = 4$: you must win, chance is 0.1
(b) $n = 5$: winning chance is $0.3 \times 0.9 = 0.27$
(c) $n = 6$: winning chance is $0.3 \times 0.6 = 0.18$
(d) $n = 7$: winning chance is $0.2 \times 0.3 = 0.06$
(e) $n = 8$: winning chance is $0.1 \times 0.1 = 0.01$

These sum to 0.62. Our table *underestimates* the skill of multiple world champion Phil 'The Power' Taylor, so at his level, 65% is a reasonable estimate of the chance the person who throws first wins the leg. With $p = 65\%$, the one who throws first wins a 3-leg set some 58% of the time, and a 5-leg set about 56% of the time.

While seeking to accumulate points towards the target of 501, top players generally aim for treble 20, the highest scoring segment on the board. Should less competent players use different tactics? The danger of aiming for 20 is that the adjacent beds score just 5 or 1; moderate players would tend to score higher if they aimed at treble 16—near misses are far less disastrous. Real tyros should go for the bull, maximising the chance of scoring something!

3.6 Tournament Design

UEFA Champions League Many soccer leagues operate on a round-robin basis; each team plays every other team home and away, and the points accumulated over the season determine the ranking. The main alternative is a knockout competition, in which teams or players are eliminated when a match is lost. In soccer, the UEFA Champions League combines these two features: 32 teams are split into 8 mini-leagues of 4 teams, each of which operates on the round-robin basis, the top 2 teams in each going on to a knockout stage.

Thus 16 teams reach this knockout stage: if the draw were purely at random, there would be $16!/(2^8 8!) = 2,027,025$ different possibilities, all equally likely. But to reward the mini-league winners, one rule is that each pairing shall consist of a Runners-up against a Champion, reducing the number of possible draws to $8! = 40,320$, again all equally likely; also, the Champions of any league shall not play the Runners-up of the same league—careful counting shows that there are now just 14,833 equally likely draws. One final rule: two teams from the same federation shall not meet, so AC Milan would not play Juventus, Barcelona would avoid Valencia, Arsenal would not meet Manchester United.

The mechanics of the draw are that eight balls representing the Runners-up are placed in a bowl, one is chosen at random: it is announced which Champions are eligible to meet that team, one of these possible opponents is selected at random in a similar fashion. But there is a danger unless precautions are taken beforehand: late in the draw, we might find that all the Champions initially eligible to play a remaining Runners-up have already been allocated other opponents, in which case the draw would collapse.

There is a neat mathematical way to avoid this fiasco: it is to use what is known as the *Marriage Theorem*. The usual setting for this Theorem is that we have a group of n women and n men; each woman has a list of those men she is prepared to marry. The Theorem asserts that, given these lists, it is possible to pair up every woman with a suitable man on her list if, and only if, for each integer r, $1 \leq r \leq n$, and each group of r women, the number of men in their joint lists is at least r. In the UEFA draw, regard the Runners-up as the women, each of which has an initial collection of eligible Champions; as the draw progresses, these lists of possible opponents may decrease in size. To ensure the draw can be completed, when a Runners-up name is drawn, UEFA's computer matches them against each ostensibly eligible remaining Champion, and if drawing that Champion would lead to a violation of the crucial condition of the Marriage Theorem for the teams that would remain, then that Champion team may not be chosen as opponents.

The matrix displayed describes the options in December 2012. Each row corresponds to a Runner-up in a mini-league, each column to the corresponding Champions. A Zero means those two teams may *not* be drawn against each other, Unity means that pairing *is* permitted. There were four Spanish teams, three German, two English, two Italian and one each from France, Portugal, Scotland, Turkey and Ukraine.

	A	B	C	D	E	F	G	H
A	0	1	1	1	1	1	1	1
B	1	0	1	1	1	1	1	0
C	1	1	0	1	0	1	1	1
D	1	1	0	0	1	1	0	1
E	1	1	1	1	0	1	1	1
F	1	1	0	1	1	0	0	1
G	1	1	1	1	1	1	0	1
H	1	1	1	1	1	1	1	0

There was some press interest in the fact that, the day before the actual draw, a rehearsal draw was held: and the real draw gave *exactly* the same set of eight pairings. How likely was it that this would occur? There is a simple formula that gives the total number of possible draws: compute what is known as the *Permanent* of the above matrix, defined by

$$\text{Perm}(A) = \sum \prod_{r=1}^{8} a_{r,\sigma(r)},$$

the sum being over all permutations σ of the numbers 1–8. (For our matrix, with entries entirely zero or one, this formula simply counts the number of paths from a 'one' in the top row to a 'one' in the bottom row, weaving the path through a series of 'ones', all in different rows and columns.)

It turns out that there were just 5463 different draws: Hans Kiesl (2013) gave an elegant proof that they are not all equally likely. He noted that, since each Runner-up has at most seven possible opponents, the chance of any particular sequence of draws is some product of the form $\Pi_{r=1}^{8}(1/k_r)$, where always $1 \leq k_r \leq 7$: a given draw may occur in many orders, so we then sum all the relevant products of this form. *If* all draws were equally likely, this sum would come to 1/5463—but the prime factors of 5463 are just 3 and 607; and since $607 > 7$, the value 1/5463 could never arise from the calculation!

The chance of the coincidence was at least 1/5463, computer simulations leading to an estimate of around 1/5400. But also, if UEFA had trusted to luck that the draw would be successfully completed without the precaution of checking the conditions of the Marriage Theorem, it would have collapsed about 22% of the time.

Using that Theorem in 2012 was essential! During the actual draw, AC Milan were the fifth Runners-up drawn; the Champions left were Malaga, Paris St-Germain, Manchester United and Barcelona. Malaga were ineligible, having won the league ahead of AC Milan, but *no random draw took place*: UEFA simply announced that AC Milan would meet Barcelona! Had they not done so, and either the French or English team been chosen as Milan's opponents, disaster would have occurred later: two remaining Spanish Runners-up would have had just one eligible opponent between them.

Chess Round-robin is a standard format in chess tournaments among $n > 2$ players; if a game is drawn, each player gets half a point, otherwise the winner scores one point, the loser zero. Of particular interest is the Sonneborn–Berger method that chess uses to break ties when two (or more) players end up with the same total number of points. Call the points total for each player their *raw score*. To break ties, a *secondary score* is computed for each player: this is found by adding up the raw scores of those players he has beaten, together with half the raw scores of those players he drew with. The idea is to suitably reward performance against the better ranked opponents.

This is neatly described using matrices. To avoid fractions, double all standard scores, so that winning gains two points, drawing gives one point; for the secondary score, add up *twice* the scores of those you have beaten, together with the scores of those you draw with. This plainly does not affect the ranking of players. As an illustration, the matrix below shows the points scored in a (fictitious) six-player tournament:

	A	B	C	D	E	F	Raw Score
A	0	2	1	1	2	1	7
B	0	0	2	1	2	2	7
C	1	0	0	1	1	2	5
D	1	1	1	0	2	0	5
E	0	0	1	0	0	2	3
F	1	0	0	2	0	0	3

Thus, on raw scores, A and B tie for first, C and D for third, E and F for fifth: the secondary scores, in order, are 33, 27, 21, 25, 11 and 17 leading to a clear ranking as $\{A, B, D, C, F, E\}$, all ties being broken. Player A obtained considerable benefit from beating B, D's draws against A and B were well rewarded in his tie-break contest with C, and F's victory against D contributed highly to him being placed ahead of E.

If M represents the matrix of scores, and $\mathbf{e} = (1, 1, 1, 1, 1, 1)^T$ is the column vector all of whose entries are unity, the raw scores are just the entries in the vector $M\mathbf{e} = \mathbf{u}$, say; and the secondary scores are the entries in $M\mathbf{u} = M^2\mathbf{e}$. This observation prompts the thought: why stop there? Why not calculate $M^3\mathbf{e}$, $M^4\mathbf{e}$ and so on? Of course, the entries in these vectors become large very quickly, but it is not the actual values, it is the *ranking* of the entries that is used to order the contestants; we could simply scale all these vectors in some sensible fashion, e.g. make them sum to unity, so divide the raw scores by 30, the secondary scores by 134, and so on.

The key mathematical idea here is the *Perron–Frobenius Theorem*, which states that, in the present context, there is some unique number $\lambda > 0$, the largest so-called *eigenvalue* of the matrix M, and a corresponding *eigenvector* $\mathbf{v}$, all of whose entries are positive, with $M\mathbf{v} = \lambda\mathbf{v}$; and the vector $M^n\mathbf{e}$, when scaled, converges to be proportional to $\mathbf{v}$. Thus we could use the entries in $\mathbf{v}$ as a neutral way of ordering the contestants, without arbitrarily stopping at some power of M. Mathematical software exists that will swiftly perform these calculations.

Just for fun, here is how one table in the 2011/12 UEFA Champions League would have worked, if the Sonneborn–Berger method had been used to break ties. The points scored across all matches were:

	A	Z	P	S	Total Points
APOEL	0	4	4	1	9
Zenit St. Petersburg	1	0	4	4	9
Porto	1	1	0	6	8
Shaktar Donetsk	4	1	0	0	5

APOEL won the league, because of the rule that, on equal points, the head-to-head matches are looked at first—and APOEL got four points, while Zenit got only one point. For Sonneborn–Berger, as extended, the secondary scores are $(73, 61, 48, 45)$, then comes $(481, 445, 404, 353)$, both leading to the same order as under head-to-head. Indeed, in the limit, the eigenvector gives the same order.

Ice-skating This event has two components; the short programme comes first, and counts just half the weight of the free programme. After the Ladies short programme in the 2002 Winter Olympic Games, the top four skaters, in rank order, were Michelle Kwan, Irina Slutskaya, Sasha Cohen, Sarah Hughes. In the free programme, Slutskaya was last to skate; before she began, the others had been ranked as Hughes, Kwan and Cohen. Temporarily ignoring Slutskaya, the provisional overall order was Kwan, Hughes, Cohen. Which of Kwan or Slutskaya will take Gold?

Neither of them! Sarah Hughes won Gold, Slutskaya Silver, Kwan took Bronze.

The explanation is the bizarre scoring system, that used just the *ranks* in each component. The rank scores from the short programme were Kwan (0.5), Slutskaya (1.0), Cohen (1.5) and Hughes (2.0), the fractions reflecting its lesser weight. In the free programme, before Slutskaya skated, the provisional rank scores were Hughes (1.0), Kwan (2.0), Cohen (3.0), giving provisional totals Kwan (2.5), Hughes (3.0) and Cohen (4.5)—the lower the better.

Had Slutskaya won the free programme, the final totals would have been Slutskaya (2.0), Kwan (3.5), Hughes (4.0), with Cohen (5.5) fourth.

Had Slutskaya finished third or lower, then the scores of Kwan and Hughes would have been the 2.5 and 3.0 noted above, with Slutskaya at least 4.0, so Kwan would win, Hughes second.

But suppose Slutskaya came second: the final totals would be Hughes (3.0), Slutskaya (3.0), Kwan (3.5), Cohen (5.5), and Hughes would win overall, because of the rule that, with a tie, the free programme dominates. And that is exactly what happened! By this method of judging, the relative ranking of Sarah Hughes and Michelle Kwan depended on how well Irina Slutskaya skated! This is not an outcome that is easy to defend.

Postscript. Thankfully, changes have been made, simple ranking is no longer used. Skaters are marked according to their success in various elements of their programme, and the scores added up. When ranking was used, it made no difference if you were first by a tiny margin, or first by a long way: now it does.

3.7   Penalty Kicks in Soccer

In taking a penalty kick, a player has many choices: to aim left, right or centre; high or low; blast or steady, etc. Similarly, the goalkeeper can stay still, dive left or right, high or low, etc. To focus on principles, and keep things simple, let each player have just *two* tactics. The Kicker, K, can aim straight, or at a corner; the Goalkeeper G can stand still, or dive one way.

For any given choice they each make, we can assess the chance a goal results. For example, if K aims straight, and G dives, the goal chance would be about 90% (K may hit it over the bar, or it might hit G's leg); but if G stays still, the goal chance will be much less. On the other hand, if K aims for the corner, he might miss the goal altogether, but if he does not, he is very likely to score if G stands still, less likely if G makes a dive. Perhaps the goal chances (%) are

	G stays still	G Dives
K aims straight	30%	90%
K aims at the corner	80%	50%

Given this table, what should K and G do? The essence of this setup is that there is a *conflict*. If K aims straight, it is best for G to stay still, if K aims at a corner, G does best to dive and vice versa. Each player would like to know what the other will do, so each must *conceal* his tactics from the other. Suppose K aims straight a proportion p of the time, at the corner otherwise. Then the chance K scores depends on what G does. Use the *columns* in the table of goal chances:

 (i) G stays still. The goal chance is $p \times 0.3 + (1 - p) \times 0.8 = 0.8 - 0.5p$.
(ii) G dives. The goal chance is $p \times 0.9 + (1 - p) \times 0.5 = 0.5 + 0.4p$.

As p increases, the former falls and the latter rises, as Fig. 3.5 shows.

Suppose we select the value of p that makes these goal chances equal, i.e. put $0.8 - 0.5p = 0.5 + 0.4p$, leading to $p = 1/3$ and the goal chance is then $X = 63.3\%$. We now have protection against any spies G has—if G knew that $p < 1/3$, G would

each use?

to the two

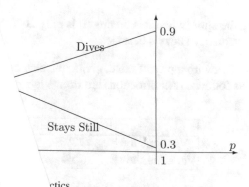

...must score if
...arry dives left.
...arry dives left,

...the overall (per-
... dives right it is
... $= 60 + 30p$, i.e.
...11 of the time, at

... equation is $72r +$
...e left 10/11 of the

...ively, if he guesses
...e in five. So his best
...miss if he goes right,
...ng the shot if Vic goes

...er extra time, and it is
...ich team goes forward
...ournament to mark the
... and Oxford University
...n gathering gloom, until
...hampionship, when Italy
..., the drawing of lots took
...in a similar fashion in the
...ecame the modern penalty

..., a Working Party for the
... penalty kicks: a coin shall
...eir opponents then take five
...the kicks alternate between
...e same number of kicks. The
...e initial kicks should also be

...ctics

'. In both cases, the chance of a goal would

rows in the table. He will stand still a
...nces of a goal are $0.3r + 0.9(1 - r) =$
...r, which are equal when $r = 4/9$. By
o more than this same value X.

context of a *two-person zero-sum*
what the other loses, but the crucial
...ard', the chance of scoring, while
...sm by increasing the number of
...ugh more complicated, will lead
...e each of the available tactics

...ended strategy? In the simple
...ave a 1/3 chance of aiming
...eings are not programmed

hows 1 or 2.
lock: aim straight if it

) is currently nearest

re if he knows you
theory, do practice

rner, either to his
—he only misses
right. Goalkeepe
t side, his chan

of saving the shot (when it is on target) is only 20%. What tactics should
Intuitively, justify Harry's best tactics.

Solution. Draw up a payoff table, giving the chance of a goal according
actions, as follows. (All directions are from Vic's perspective.)

	Harry dives left	Harry dives right
Vic aims left	72%	90%
Vic aims right	75%	60%

Justification: if Vic aims left, he is on target 90% of the time, so
Harry dives right, and will score $80\% \times 90\% = 72\%$ of the time if H
If Vic aims right, he is on target 75% of the time, so then scores if F
and still scores $80\% \times 75\% = 60\%$ of the time if Harry dives right.
Suppose Vic aims left with probability p. If Harry dives left,
centage) chance of a goal is $72p + 75(1 - p) = 75 - 3p$, if he
$90p + 60(1 - p) = 60 + 30p$. These are equal when $75 - 3p$
when $p = 5/11$. Vic should aim left 5/11 of the time, and right 6
random.
For Harry, if he dives left with probability r, the correspondin
$90(1 - r) = 75r + 60(1 - r)$, i.e. $r = 10/11$. Harry should di
time, and right just 1/11 of the time, again at random.
Why should Harry overwhelmingly dive to the left? Intui
correctly, and the shot is on target, he still only stops it one tin
chance of 'no goal' is for Vic to miss. Vic is far more likely to
so Harry should concentrate on increasing his chance of stoppi
left. Hence the 10/11.
In knockout competitions, if the scores are level even af
not practicable to organise a replay, some way of deciding w
is needed. Various methods have been used: in 1963, in a
centenary of the Football Association, Sheffield Universit
were level at the end of extra time: the match continued
Oxford scored after 152 min in all. In the 1968 European (
and the Soviet Union had scored no goals at all in 120 mi
Italy to the final. After Israel was eliminated by Bulgaria
1968 Olympic Games, Yosef Dagan proposed what soon
shoot-out.
In 1970, although not wholly convinced of its mer
International Board recommended a method based on
be tossed, the winners take five consecutive kicks, th
kicks. At level scores, 'sudden death' should apply—
the teams until one is ahead when both have taken th
International Board accepted this, except that the fi
taken alternately—and that still applies.

However, after 40 years, enough data had accumulated to show clearly that the team who won the toss and took the first kick had a significant advantage: in 1001 shoot-outs, with 10,431 kicks, the team kicking first won 60% of the time. The method used (teams simply alternating) could be described as ABABAB...; if kicks are scored (and this happens 70 to 80 percent of the time), team B is always playing catch-up, seen as a psychological disadvantage. To overcome this, the 'ABBA' order, ABBA ABBA ABBA... has been suggested—this is the rule followed when tennis players play tie-breaks. After 1, 5, 9, 13,...shots, team A has taken more kicks, while team B has taken more after 3, 7, 11, ... Another suggestion is to use the so-called Thue–Morse order, i.e. ABBA BAAB ABBA BAAB ... . Steven Brahms and Mehmet Ismael have proposed the 'catch-up rule': in any 'round' of two kicks, each team takes one of them, a coin-toss deciding who takes the first kick. But subsequently, the first kick is taken by the team that *lost* in the previous round. (If the previous round were drawn, the order alternates.) But might players or referees get confused if a system is too elaborate?

3.8 Golf: Flamboyance Versus Consistency

In golf, a risky shot may gain, but could also prove costly. For given overall ability, will a steady consistent player tend to do better or worse than a flamboyant competitor who takes risks?

A central idea in golf is 'par': most holes are 'par 4', with a competent player expected to take two shots to land the ball on the green, one putt to get near the hole, one more putt to sink it. Shorter holes are 'par 3', longer ones may be 'par 5'. So, from any position on the course, there is some 'par score' expected to complete the hole: close to the hole, it will be 'par 1', six metres away 'par 2', etc.

For a simple model to explore this idea, we suppose there are three types of shot: an *Ordinary* shot does what is expected for par, a *Good* shot saves one shot to par, a *Bad* shot essentially leaves the ball where it is, losing one shot to par. We assume the player is equally likely to hit a Good shot as a Bad one so that, on average, all shots are Ordinary. Good and Bad shots each occur a proportion x of the time, so that Ordinary shots will happen $1 - 2x$ of the time. How does the average number of shots for a hole vary with x?

Write A_r as the average number of shots it will take when at a point whose distance from the hole would be 'par r', for $r = 1, 2, 3, \ldots$. When $r = 1$, the ball is holed with frequency $1 - x$, but a proportion x of the time, it effectively stays where it is. So we have

$$A_1 = 1 + xA_1,$$

leading to the simple result $A_1 = 1/(1 - x)$.

For $r \geq 2$, the next shot either leaves the ball where it is (frequency x), moves it one par shot closer (frequency $1 - 2x$), or two par shots closer (frequency x again), giving the recurrence relation

$$A_r = 1 + x.A_r + (1 - 2x).A_{r-1} + x.A_{r-2}.$$

We can use this to find the particular values $A_2 = (2 - 3x)/(1 - x)^2$, $A_3 = (3 - 8x + 6x^2)/(1 - x)^3$, etc., with the general result (Exercise 3.19)

$$A_r = r + x + (-1)^{r-1}\frac{x^{r+1}}{(1 - x)^r}.$$

So when r is an odd number, A_r always increases with x; when r is even, this also holds for the fairly small values of x seen in practice. Thus, as a general rule, a more consistent golfer (lower x) will take fewer shots, on average, than a flamboyant risk-taker.

But a flamboyant player may well win more money. Prizes are skewed towards the winner, and our consistent player may finish about 20th every time, while the flamboyant player finishes 30th *on average*—but the prize money coming from the occasional top-five finishes could outweigh the steady moderate amounts won by the consistent player.

3.9   Exercises

3.1 We have argued that, if p is the chance of winning a point in tennis, then when $p \neq 1/2$, the chance of winning the game can be written as

$$G(p) = \frac{p^4 - 16p^4(1 - p)^4}{p^4 - (1 - p)^4}.$$

Use L'Hôpital's Rule to show that, as $p \to 1/2$, so also $G(p) \to 1/2$.
Let $p = 0.5 + \epsilon$, where $\epsilon > 0$ is small. Show that G is approximately $0.5 + 2.5\epsilon$, and compare the correct value of G with this approximation when $p = 52\%$ and $p = 60\%$.
Figure 3.1 shows the graph of G against p. If you already win 80% of points on your serve, how many more service games would you expect to win if you could increase this to 90%?

3.2 In 2010, John Isner beat Nicolas Mahut 70–68 in the fifth set in the Wimbledon tournament. (Both players have powerful serves.) To reduce the chance of such a long set, it has been suggested that each game might begin with the score already 0–15. Adapt the argument in this chapter to show that, with such a rule change, the chance the server wins the game (if $p \neq 1/2$, and $q = 1 - p$) would be

$$H(p) = p^4 + 4p^4q + 10p^3q^2.\frac{p^2}{1 - 2pq} = \frac{p^4(1 - 10q^3 + 4q^4)}{p^4 - q^4}.$$

For a server who wins 80% of the points on serve, by how much would this rule change decrease the chance the server wins the game?

3.3 Let F(S) denote the Fast(Steady) serve of a tennis player; let x, y be the respective chances that F(S) are good (not faults), and f, s be the respective chances of winning the point if the two types are good.

Take $f = 80\%$, $s = 50\%$, $x = 50\%$ and $y = 90\%$. For each of FF, FS, SF, SS find the winning chances; rank these tactics.

Invent *plausible* values (i.e. with $f > s$ and with $y > x$) for f, s, x, y making: (i) FF best;   (ii) SS best.

3.4 In the notation of this chapter, *prove* the claims that, when $f > s$ and $y > x$, then FF has the largest chance of winning a point if $R > 1$, while SS is best when $R < 1 + x - y$ and FS is best otherwise.

3.5 Suppose that the chance Alice wins any *set* in tennis against Betty is p ($0 < p < 1$), irrespective of who serves first and the results of other sets. The winner of the match is the first to win two sets. List the ways in which Alice could win the match, and hence deduce that the chance she does so is $p^2(3 - 2p)$. Deduce that, if $p > 0.5$, Alice is more likely to win the match than to win the first set.

3.6 In lawn tennis, if the set score reaches 6–6, a *tie-break* may be played. The first player to win seven points, and be at least two ahead of his opponent, wins the set 7–6. Suppose Ted beats Harry 0–6, 0–6, 7–6, 7–6, 7–6: show that it is possible for Harry to have won some 65% of the points.

3.7 In table tennis, estimate the total number of points played in a match between two equally good players under the old rules, and the new (see the text at the end of the Lawn Tennis section).

3.8 In the notation of Fig. 3.2, when seeking to convert a try scored at the point X (wide of the posts), the angle made by the posts (C, D) and the conversion point Y is maximised when XY is a tangent to the circle through C, D and Y. Choose a coordinate system with the x−axis along CD, the origin being at the centre of the goal and the y−axis parallel to XY. Take C as $(-a, 0)$, D as $(a, 0)$, and Y as (x, y), with angle CYD $= \theta$.

Write θ as the difference between two angles to show that

$$\tan(\theta) = \frac{2ay}{(x^2 + y^2 - a^2)}.$$

Since maximising θ is the same as maximising $\tan(\theta)$ (why?), find the optimal value of y, given x, thereby showing that the locus of the optimal points X satisfies $x^2 - y^2 = a^2$ (a *rectangular hyperbola*).

3.9 Two snooker balls each have radius $r > 0$, their centres are distance $2xr$ apart with $x > 1$. If the cue ball is sent at an angle θ away from the line of the centres, but small enough to hit the object ball, we have shown that the object ball is deflected at an angle ϕ, where

$$\tan(\theta) = 2r \sin(\phi)/(2xr - 2r \cos(\phi)).$$

Deduce that $\phi = \arcsin(x \sin(\theta)) - \theta$; hence give the actual values of ϕ when $x = 10$ and θ is (i) one-tenth of a degree (ii) one degree.

In the rest of this question, the centre of the cue ball is distance $2Dr$ from the pocket, all other balls are placed in the straight line between the cue ball and the centre of the pocket, the cue ball is hit hard and $\theta \neq 0$.

(a) Suppose there is one other ball, distance $2xr$ from the cue ball with $1 < x < D$. If θ is small and the cue ball hits the object ball, show that the object ball will be approximately $2r(x - 1)(D - x)\theta$ away from the pocket's centre. What value of x maximises the distance of the object ball from the pocket? What is this distance (approximately)?

(b) Suppose *two* other balls are placed, the first at distance $2xr$ from the cue ball, the second at a further distance $2yr$ form the first, with $x > 1$, $y > 1$ and $x + y < D$. If θ is small enough for the cue ball to propel the first object ball into the second, show that the second ball will miss the pocket by $2r(x - 1)(y - 1)(D - x - y)\theta$, approximately. Deduce that this quantity is maximised when $x = y = (D + 1)/3$, and write down the corresponding approximation to the distance by which the second ball misses the pocket.

3.10 When $a > b > 0$, the relation

$$\frac{x^2}{a^2} + \frac{y^2}{b^2} = 1$$

describes an *ellipse* with *foci* at $S = (-ae, 0)$ and $H = (ae, 0)$, where $e > 0$ is its *eccentricity*, given by $a^2 e^2 = a^2 - b^2$. Let $P = (x_1, y_1)$ be an arbitrary point on this ellipse.

  (i) Use the hint in the text, and the fact that tangent and normal are orthogonal, to show that $\mathbf{n} = (b^2 x_1, a^2 y_1)$ is in the direction of the normal at P.
 (ii) Let θ and ϕ be the angles between this normal and the lines SP and HP, respectively, joining P to the two foci. Use scalar products to show that $\cos(\theta) = \cos(\phi) = ab^2/|\mathbf{n}|$.
(iii) Place the cue ball at one focus of such an elliptical billiard table (with no pockets) and cushions that give true rebounds. Describe its path, when hit in any direction, but so that the total distance it travels is $4a$.

3.11 A shot-putter releases the shot at V metres/second, from height h metres, and at an angle θ above the horizontal, and g m/sec^2 is the acceleration due to gravity. Use the formula for the distance achieved to show that the optimal value of θ comes from $\sin^2(\theta) = 0.5V^2/(V^2 + gh)$. Take $V = 14$, $h = 2$ and $g = 10$ (plausible for top-class male shot-putters), and verify that, with these figures, the best value of θ is around $42°$. Give the maximum distance achieved (to the nearest $5\,\mathrm{cm}$).

Take $\theta = 42°$, and V, h, g as above. How much further (to the nearest cm or so) would the shot travel, with (three separate answers) (i) a 1% increase in V (ii) a 1% increase in h (iii) a 1% decrease in g?

According to the National Physical Laboratory, the formula for g at latitude ϕ and altitude A metres above sea level is (very nearly)

$$g = 9.7803(1 + 0.0053\sin^2(\phi) - 0.000006\sin^2(2\phi)) - 0.0000031A.$$

Estimate the extra distance (cm) a top-class male shot-putter would achieve in Mexico City (latitude 19°, altitude 2300 m) as opposed to sea level Helsinki (60° North).

3.12 Cricketer Kevin practises straight hits. The ball leaves his bat at 30 m/s, launched at an angle of *either* 60, *or* 45, degrees above the horizontal. Joe stands directly in the line of the hit, distance D metres from Kevin, and will try to catch the ball (at exactly the same height above ground as where it leaves the bat). Joe will face Kevin all the time, and can run at 5 m/s forwards, or 3 m/s backwards. Find the largest value of D so that Joe can reach the ball running forwards if it is hit at the steeper angle, and the smallest value of D so that he can reach the ball running backwards, if it is hit at the shallower angle. (Take $g = 10$ m/s/s.)

3.13 In the decathlon, the score for a High Jump of x cm is $0.8465(x - 75)^{1.42}$, provided $x \geq 75$. Use the Taylor expansion

$$f(x + h) \approx f(x) + hf'(x)$$

to estimate the extra points scored for each extra increment of two cm achieved, for an athlete currently jumping around 2.20 m.

3.14 (i) In how many ways can we arrange the numbers 1–20 on a dartboard, with odd and even numbers alternating?

(ii) In the notation of this chapter, prove that $S \leq 200$ for all arrangements of $\{1, 2, \cdots, 20\}$ round the board.

(iii) The 'Low' numbers are 1–10, the 'High' numbers are 11–20. *Prove* that it is not possible to have an arrangement where *both* High and Low numbers alternate, *and* Odd and Even numbers alternate.

(iv) A darts match is set up as the best of five sets, each set having five legs. Use the appropriate text formula (twice) to find the chance that the player who throws first in the first leg of the first set wins this match, using the value $p = 0.7$. (They alternate the right to go first from set to set.)

(v) Tina and Kate are equally good at darts; whoever throws first wins a standard game with probability $p > 0.5$. They alternate who goes first, and use an unusual rule: the winner is the first to win two *consecutive* games. Tina can decide who throws first in the first game: should she choose to go first, let Kate go first, or does it make no difference?
(Assume that the outcomes of different games are independent, meaning that the overall chance of a particular list of outcomes arises by multiplying together the chances of the outcomes of the individual games.)

3.15 Suppose that, during a draw for the UEFA Champions League, using the rules given in the text, the position when 12 teams are left is as follows. (The Runners-up are A to F, the Champions are P to U; we list the Champions eligible to play the given Runners-up.) A: QRT; B: QT; C: PQRSU; D: RST; E: QRT; F: PQSTU. Verify that the draw can be legally completed. Now suppose that team D are the next Runners-up selected: describe what will happen.

3.16 Suppose that, in a given year, there are N valid draws for the UEFA Champions League, with respective probabilities $\{x_1, x_2, \ldots, x_N\}$. Show that the chance that the rehearsal draw and the real draw give identical fixtures is $\sum x_i^2$. Prove that the value of this expression is at least $1/N$.

3.17 In a chess tournament with four players, A beats B and C but loses to D; B beats C and D; C beats D. How does the Sonneborn–Berger method resolve the ties for first and third places? Write M as the matrix of results, giving one point for a win, zero for a loss and let $\mathbf{e} = (1, 1, 1, 1)^T$; compute the successive values of $M^r\mathbf{e}$ for $r = 1, 2, 3, \ldots$ until all components of the vector are different. What overall order would be vector imply?

3.18 We have set up the taking of a penalty kick as a conflict between the Kicker K and the Goalkeeper G. The chances of a goal, according to which of their two tactics they use, have the form

	G stays still	G dives
K aims straight	a	b
K aims at the corner	c	d

with $a < b, c > d, a < c$ and $b > d$. For each of G and K, find (in terms of a, b, c, d) the proportion of the time they should use each of their tactics, in order to guard against the possibility that the other has spies who may discover this.

If these proportions are used, what is the overall chance of a goal?

3.19 Use induction to prove the general formula

$$A_r = r + x + (-1)^{r-1}\frac{x^{r+1}}{(1-x)^r}$$

for the mean number of shots to complete a hole from a point which is 'par-r' in the golf model described in the text.

3.20 An 18-hole golf course has three par-3 holes, three par-5, the rest being par-4. For the golf model described in the text, what is the average score for one complete round for a player who hits Good or Bad shots each with frequency 5%, Ordinary shots with frequency 90%?

References and Further Reading

Bennett J (1998) Statistics in Sport. Arnold

Brahms S J and Ismael M (2018) Making the Rules of Sports Fairer *SIAM Review* 2018 60(1) pages 181–202

Butenko S, Gil-Lafuente J and Pardalos P M (2004) Economics, Management and Optimization in Sports. Springer

Eastaway R and Haigh J (2011) The Hidden Mathematics of Sport. Portico

Haigh J (2003) Taking Chances. Winning with Probability. OUP

Haigh J (2009) Uses and Limitations of Mathematics in Sport. *IMA Journal of Management Mathematics* 2009 20(2) pages 97–108

Kiesl H (2013) Match me if you can. *Mitteilungen der DMV* 21 pages 84–8

Morris C (1977) The most important point in tennis. In "Optimal Strategies in Sports" edited by S P Lahany and R E Machol. Elsevier

Palacios-Huerta I (2014) Beautiful game theory: how soccer can help economics. Princeton University Press

Percy D (2012) The Optimal Dartboard? *Mathematics Today* 48(6) pages 268–70

Singmaster D (1980) Arranging a dartboard. *IMA Bulletin* 16(4) pages 93–7

Wallace M and Haigh J (2013) Football and marriage – and the UEFA draw. *Significance* 10(2) pages 47–8

Chapter 4
Business Applications

4.1 Stock Control

Suppose a retailer sells non-perishable non-seasonal goods, such as kitchen equipment, car accessories or cleaning necessities. The estimate of total annual demand is N items, which must be obtained from some wholesaler, and stored until sold: thus costs comprise

(i) a fixed amount C for each order, including delivery;
(ii) the wholesale cost of each item, P;
(iii) storage costs, of H per item per year.

If demand is steady, how many should be ordered each time so as to minimise overall annual costs?

Ordering Q at a time implies N/Q orders per year in the long run, at an annual order cost of CN/Q. Buying the items costs NP, and with steady demand, the average number in storage at any time (above an irreducible minimum) will be $Q/2$: so we assess the annual storage costs as $HQ/2$. In total, the average annual cost is

$$f(Q) = NP + CN/Q + HQ/2.$$

To find the value of Q that minimises this, put the derivative of $f(Q)$ to zero, i.e. solve

$$-CN/Q^2 + H/2 = 0,$$

leading to $Q^2 = 2CN/H$.

Since $f''(Q) = 2CN/Q^3 > 0$, this does indeed give a minimum. The quantity $\widehat{Q} = \sqrt{2CN/H}$ that minimises these long-term costs is termed the *Economic Ordering Quantity* (EOQ).

© Springer Nature Switzerland AG 2019
J. Haigh, *Mathematics in Everyday Life*,
https://doi.org/10.1007/978-3-030-33087-3_4

Fig. 4.1 Variation of annual
cost with size of each order

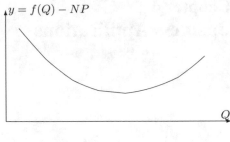

Fig. 4.2 Annual cost, with
discounts for bulk orders

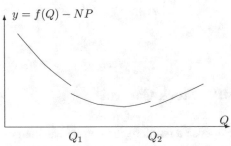

The function $f(Q)$ is the sum of a constant (NP), a decreasing function (CN/Q) and an increasing one ($HQ/2$). The general shape of

$$y = f(Q) - NP$$

is shown in Fig. 4.1.

A supplier might give discounts on sufficiently large orders, perhaps 5% on orders of at least Q_1, and 10% on orders of at least Q_2. This has no impact on storage or delivery costs, but there will be a step change in the overall cost at any point where the discount kicks in: for any quantity Q where the discount is of size θ, total cost drops by θNP. Figure 4.2 shows the shape of the new graph of $y = f(Q) - NP$, if there is a discount on orders of at least Q_1, and a higher discount for orders of size at least Q_2.

For any value Q where the original graph is strictly *decreasing* (e.g. at the point Q_1 here) this discount point cannot be a global minimum, as a marginally higher Q leads to a lower total cost. But where the original graph is increasing or level, a discount might well (as at the point Q_2 here) be a global minimum. Thus, we just have a finite list of candidates for the minimum possible cost—the original $\widehat{Q}$ already identified, and those discount points where $f'(Q) \geq 0$.

Example 4.1 Megastore sells electric toasters, and estimates annual demand at 1000 items. Their wholesale suppliers normally charge £10 for each such toaster; it costs Megastore £20 to place an order (including delivery), and they take storage costs as £4 per item per year. What order size minimises costs for Megastore? If the suppliers give a discount of 2% on order sizes of at least 50, 3% on order sizes of at least 100, and 5% if the order is for 200 or more, what order size would now minimise costs?

Solution. In our notation, we have $N = 1000$, $P = 10$, $C = 20$ and $H = 4$, and it seems reasonable to assume that demand is steady. So if Megastore order Q at a time, their annual cost (in £) will be

$$f(Q) = NP + CN/Q + HQ/2 = 10{,}000 + 20{,}000/Q + 2Q,$$

with an EOQ of $\sqrt{2CN/H} = \sqrt{40 \times 1000/4} = 100$. The store should order 100 items each time, if no discounts are offered.

Since the threshold of 50 for the 2% discount is below this quantity, this cannot yield a global minimum. We must check on the other two possibilities:

(i) when $Q = 100$, the new cost (above 10,000) is

$$20{,}000/100 + 2 \times 100 - 10{,}000 \times 3\% = 200 + 200 - 300 = 100;$$

(ii) when $Q = 200$, the new cost (above 10,000) is

$$20{,}000/200 + 2 \times 200 - 10{,}000 \times 5\% = 100 + 400 - 500 = 0.$$

With such discounts, they should order 200 at a time.

But consider the problem of stocking goods with a short shelf life, such as pork pies or daily newspapers. Suppose a newsagent believes that the probability the demand for tomorrow's newspaper will be n, with probability p_n for $n = 0, 1, 2, 3, \ldots$ If she orders too many, she faces a loss of L units for each unsold copy, but if she orders too few, each extra copy she could have sold reduces her profits by P units. How many should she order?

Write $S_Q = \sum_{n=0}^{Q} p_n$ as the chance that demand turns out to be *at most* Q, and consider the difference between ordering $Q + 1$ rather than Q. The chance that the extra copy will be left unsold is S_Q, leading to a loss of size L; since the chance that demand is *more than* Q is $1 - S_Q$, the extra copy ordered and sold leads to a profit of P. On average, the difference between ordering $Q + 1$ rather than Q is

$$P(1 - S_Q) - L S_Q = P - (L + P) S_Q.$$

This is positive so long as $S_Q < P/(L + P)$, so the optimal decision is to order the smallest amount Q that makes $S_Q \geq P/(L + P)$.

Example 4.2 Suppose each copy left unsold would cost 90 pence, while the profit on each copy sold is 20 pence, and the demand is taken to be somewhere in the range from 81 to 100 copies, all judged equally likely. Then, in our notation, $L = 90$, $P = 20$, so $P/(L + P) = 20/110 = 2/11$. Plainly, $S_Q = 0$ for $Q \leq 80$, $S_Q = 1$ for $Q \geq 100$, and S_Q rises linearly in the range from 81 to 100, so takes the value $(Q - 80)/20$ at integer points. Now $(Q - 80)/20 \geq 2/11$ so long as $Q \geq 83.636363\ldots$, so she should order 84 copies to maximise her expected profit.

4.2   Linear Programming

In this introduction, we will merely describe what this subject is, and how to solve the problems it throws up. Precisely *why* the methods work is not covered. And we will look only at examples that involve only a small number of variables, although, in applications, the number of variables can be in the tens of thousands, or even more.

Suppose a firm supplying m different food items must produce at least (say) b_1 tons of sugar, b_2 litres of apple juice, b_3 kilograms of muesli, etc., as cheaply as possible. It has n production sites, with different costs and capabilities: to run site j for one day costs c_j, and it will produce a_{ij} units of item i, for $i = 1, 2, \ldots, m$. Suppose site j runs for x_j days.

To meet demand, we must have $a_{i1}x_1 + a_{i2}x_2 + \cdots + a_{in}x_n \geq b_i$ for each value of $i = 1, 2, \ldots, m$. The total cost will be $c_1x_1 + c_2x_2 + \cdots + c_nx_n$. So we seek to

$$\text{Minimise } c_1x_1 + c_2x_2 + \cdots + c_nx_n, \text{ subject to } a_{i1}x_1 + a_{i2}x_2 + \cdots + a_{in}x_n \geq b_i$$
$$\text{for each } i = 1, 2, \ldots m, \text{ with each } x_i \geq 0.$$

In vector and matrix notation: let $\mathbf{c}' = (c_1, c_2, \ldots, c_n)$ be a row vector, and let $\mathbf{x} = (x_1, x_2, \ldots, x_n)'$, $\mathbf{b} = (b_1, b_2, \ldots, b_m)'$ be column vectors. Take A as the $m \times n$ matrix (a_{ij}). The problem can be expressed as

$$\text{Minimise } \mathbf{c}'\mathbf{x} \text{ subject to } A\mathbf{x} \geq \mathbf{b} \text{ and } \mathbf{x} \geq \mathbf{0}.$$

This is called the *Standard Minimum Problem* of linear programming. The quantity of interest, $\mathbf{c}'\mathbf{x}$, is termed the *objective function*.

There is a logical method to solve such problems, which rests on a simple idea. (To visualise matters, think of n being 2 or 3, so you have a mental picture of the (x, y) plane, or ordinary 3-space.) Equations like

$$x_i = 0 \quad \text{or} \quad a_{i1}x_1 + a_{i2}x_2 + \cdots + a_{in}x_n = b_i$$

define lines or planes that split the whole space into two halves. An inequality such as $x_i \geq 0$ or $a_{i1}x_1 + a_{i2}x_2 + \cdots + a_{in}x_n \geq b_i$ specifies the points that lie on one side of this line or plane. (In higher dimensions, we refer to 'hyperplanes'.) To satisfy all such constraints, we want the *intersection* of the half-spaces defined by these hyperplanes. Because each $x_i \geq 0$, we are interested only in points in the positive quadrant/octant/higher dimensional analogue. The points where all the inequalities hold is called the *feasible region*.

This feasible region will always be a *convex polytope*: in two dimensions, the edges are straight lines, in 3-space, all its faces are planes, all its edges straight lines. It will generally contain infinitely many points, and the minimum of an *arbitrary* function $f(\mathbf{x})$ could be anywhere in the feasible region. *But* when the function takes the *linear* form $\mathbf{c}'\mathbf{x}$, the minimum can only occur at one of the *vertices* of the polytope—

Fig. 4.3 The feasible region, and the objective function

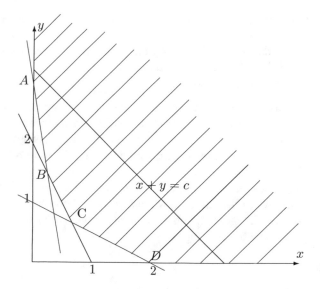

of which there are only finitely many! Example 4.3 illustrates why. (With just two variables, we use x, y rather than x_1, x_2.)

Example 4.3 Minimise $x + y$ subject to

$$2x + y \geq 2, \quad x + 2y \geq 2, \quad 6x + y \geq 3 \text{ and } x \geq 0, y \geq 0.$$

Solution. The feasible region, whose vertices are $A = (0, 3)$, $B = (1/4, 3/2)$, $C = (2/3, 2/3)$ and $D = (2, 0)$, is shaded in Fig. 4.3. We draw a straight line $x + y = c$, seeking for c to be as small as possible, consistent with some part of the line falling in the feasible region. As c reduces, the line moves parallel to itself towards the origin, so that *any such line can leave the feasible region ONLY at the vertices!*

So to solve the problem, just work out the values of the objective function at the vertices, and see which is smallest. At A, B, C and D, the respective values of the objective function are $3, 7/4, 4/3$ and 2, the smallest of which is $4/3$ achieved at point $C = (2/3, 2/3)$ (as is apparent from the graph). If the objective function had been $5x + y$, with the same feasible region, the minimum value, $11/4$, would arise at $B = (1/4, 3/2)$.

Exercise 4.3 asks you to minimise $3x + y$, and $x + 4y$, under the same constraints. So you need to do no more than evaluating each objective function at the four vertices, to identify where the minimum lies.

Reducing a search from infinitely many points to maybe 4 or 5 is a major advance. The *Simplex Method* is a logical way to solve such problems, even with thousands of variables. It consists of

(a) locating ONE of these vertices;

(b) using a criterion that checks whether this is the one that minimises the objective function;
(c) if that vertex does not yield a minimum, then locating a vertex with a smaller value of the objective function;
(d) returning to step (b).

The details can be found in any standard textbook on Linear programming. It was developed by George Dantzig in 1947, in response to logistical problems in the deployment of forces and equipment in the US Air Force. One of the first practical applications was to construct an adequate diet at minimum cost—see Exercise 4.5 (iii).

Sometimes, there is one other practical constraint: that the answer only makes sense when the values x_i are *integers*. In such cases, if the methods described lead to a non-integer solution, it is usually enough to look at the nearest all-integer points within the feasible region.

4.3   Transporting Goods

Given m sources where goods are currently stored, and n destinations where we would like them to be, how do we find the cheapest way of getting the right quantities of goods to their intended locations? This is known as the *transportation problem* and, in many instances, is also a problem amenable to techniques developed in the general theory of linear programming.

Suppose the cost of moving one item from the ith source to the jth destination is $c_{ij} \geq 0$, and we will move $x_{ij} \geq 0$ units from source i to destination j. Then the cost, which we seek to minimise, is expressed as

$$\sum c_{ij} x_{ij}.$$

If there are s_i items at the ith source, we must have

$$\sum_j x_{ij} \leq s_i;$$

and if at least d_j items are required at the jth destination, then also

$$\sum_i x_{ij} \geq d_j.$$

Plainly, for any solution to exist, the total supply of goods must be at least as great as the number required at the destinations and, if necessary, by introducing an extra destination which will receive any surplus goods, at zero transportation cost, we can assume that supply and demand exactly balance out. This lets us formulate the problem in the following standard way: given quantities $\{s_i\}$ and $\{d_j\}$, $s_i > 0$, each $d_j > 0$ and

$$\sum_{i=1}^{m} s_i = \sum_{j=1}^{n} d_j,$$

we seek to minimise

$$\sum_{i=1}^{m} \sum_{j=1}^{n} c_{ij} x_{ij}$$

subject to $\sum_{j=1}^{n} x_{ij} = s_i$, $\quad \sum_{i=1}^{m} x_{ij} = d_j \quad$ and each $x_{ij} \geq 0$.

We describe a way of solving this problem, and outline why it works later, but a formal justification must be found elsewhere. The general idea is to start with a *feasible* solution, i.e. some method of getting the right quantities of goods to their destinations, checking whether it can be bettered and, if so finding an improvement—exactly the template noted above for the Simplex Method. We have set up the problem ensuring that $\sum s_i = \sum d_j$; there is a slight awkwardness if these numbers are such that a *partial* sum of the values $\{s_i\}$ is equal to a partial sum of the $\{d_j\}$, so we will initially assume this is not the case.

First, set up a framework that summarises the problem, showing the supplies, demands and costs in a rectangular array, and introducing quantities u_i and v_j that will lead to a solution.

$$
\begin{array}{cc|ccccc}
 & d & d_1 & d_2 & \cdots & d_n & \\
 s & u_i & v_j & & \cdots & & \\
\hline
 s_1 & & c_{11} & c_{12} & \cdots & c_{1n} & \\
 s_2 & & c_{21} & c_{22} & \cdots & c_{2n} & \\
 \vdots & & \vdots & \vdots & & \vdots & \\
 s_m & & c_{m1} & c_{m2} & \cdots & c_{mn} & \\
\end{array}
$$

Example 4.4 Suppose we have $m = 2$ supply points holding 4 and 7 units, respectively, with demands $5, 3, 3$ at $n = 3$ destinations. Costs are $c_{11} = 6, c_{12} = 5, c_{13} = 8$, $c_{21} = 2, c_{22} = 3$ and $c_{23} = 1$. Then the initial layout is

$$
\begin{array}{cc|ccc}
 & d & 5 & 3 & 3 \\
 s & u_i & v_j & & \\
\hline
 4 & & 6 & 5 & 8 \\
 7 & & 2 & 3 & 1 \\
\end{array}
$$

We need an initial feasible solution, i.e. a suitable $m \times n$ matrix $\{x_{ij}\}$ which describes some way of moving the supplies to their destinations. One way is the corner method': take x_{11} as the smaller of the quantities s_1 and d_1: if we have not yet met the demand at the first destination, so choose x_{21} ~e as possible without exceeding the smaller of s_2 and the residual demand; ~reas if $s_1 > d_1$, choose x_{12} as large as possible without exceeding the smaller of ~e residual supply, or d_2. Continue in this fashion, moving from left to right and top to bottom of the matrix. Our assumptions ensure that we can always take the next desired step. Insert the non-zero values $\{x_{ij}\}$ *after* the corresponding costs. There will automatically be $m + n - 1$ non-zero values of the $\{x_{ij}\}$, and this property (a 'basic' feasible solution) must be maintained in the subsequent steps. Here, this path would lead to

$$
\begin{array}{cc}
d & 5 \quad\quad 3 \quad\quad 3 \\
s\ u_i \quad v_j & .\quad\quad .\quad\quad .
\end{array}
$$

$$
\begin{array}{cc}
4\ . \\
7\ .
\end{array}
\left(
\begin{array}{ccc}
6 \ 4\ |\ 5\ .\ |\ 8\ . \\
2\ 1\ |\ 3\ 3\ |\ 1\ 3
\end{array}
\right)
$$

Work on this array, using the non-zero values of x_{ij}. Set $u_1 = 0$, and find all the other values of u_i and v_j in turn from the equations $v_j - u_i = c_{ij}$ *whenever* $x_{ij} > 0$. Then calculate the values of $v_j - u_i$ in the remaining cells, where $x_{ij} = 0$—mark them with an asterisk. Here, we now obtain

$$
\begin{array}{cc}
d & 5 \quad\quad 3 \quad\quad 3 \\
s\ u_i \quad v_j & 6 \quad\quad 7 \quad\quad 5
\end{array}
$$

$$
\begin{array}{cc}
4\ 0 \\
7\ 4
\end{array}
\left(
\begin{array}{ccc}
6\ 4\ |\ 5\ 7^*\ |\ 8\ 5^* \\
2\ 1\ |\ 3\ 3\ |\ 1\ 3
\end{array}
\right)
$$

If all the asterisked values are less than or equal to the adjacent costs, we have found an optimal solution; but if some asterisked value exceeds the cost c_{ij}, we can find a better solution that sends as much as possible down that route. In the example, since $v_2 - u_1 = 7 > 5 = c_{12}$, we should send some amount, θ say, from supply 1 to destination 2. This will require modifications to the previous values of the $x's$, some bigger, some less, to satisfy the constraints. This notion leads to

$$
\begin{array}{cc}
d & 5 \quad\quad 3 \quad\quad 3 \\
s\ u_i \quad v_j & 6 \quad\quad 7 \quad\quad 5
\end{array}
$$

$$
\begin{array}{cc}
4\ 0 \\
7\ 4
\end{array}
\left(
\begin{array}{ccc}
6\ 4-\theta\ |\ 5\ \ \theta\ \ |\ 8\ . \\
2\ 1+\theta\ |\ 3\ 3-\theta\ |\ 1\ 3
\end{array}
\right)
$$

The largest value of θ that keeps all the x_{ij} non-negative is $\theta = 3$, so the new display, with re-calculated values of the u_i and v_j is

$$
\begin{array}{cc}
d & 5 \quad\quad 3 \quad\quad 3 \\
s\ u_i \quad v_j & 6 \quad\quad 5 \quad\quad 5
\end{array}
$$

$$
\begin{matrix}
4 & 0 \\
7 & 4
\end{matrix}
\qquad
\left(
\begin{array}{cc|cc|cc}
6 & 1 & 5 & 3 & 8 & 5^* \\
2 & 4 & 3 & 1^* & 1 & 3
\end{array}
\right)
$$

This time, the two asterisked values are indeed less than or equal to the corresponding costs, so we have found an optimal solution: $x_{11} = 1, x_{12} = 3, x_{21} = 4$ and $x_{23} = 3$. The total cost is 32, the cost of the initial feasible solution would have been 38.

To see why this procedure leads to an optimal solution, imagine that the whole job of transporting the goods is put out to tender. A firm applying for the tender will buy all the goods at the supply points, and sell them in the correct quantities at the demand points. The important factor is not the prices themselves, but just the *difference* between buying and selling costs. So we can arbitrarily set $u_1 = 0$, i.e. the goods at the first supply point are acquired at zero cost. The equations $v_j - u_i = c_{ij}$ whenever $x_{ij} > 0$ then fix all the other buying costs of the supplies, and selling costs at the demand points. But if any asterisked value—the difference between buying and selling costs for a route that this tendering firm will *not* use—exceeds the corresponding c_{ij}, the tender will be unsuccessful, as the original firm could do the job cheaper using that route just identified. Only when no asterisked value exceeds the corresponding cost can the tendering firm match the minimum overall cost that can be achieved.

A possibly better way of beginning (in that it may well require fewer subsequent steps) is the 'matrix minimum' method: locate the smallest cost among all the $\{c_{ij}\}$, and send the largest possible amount along that route; go to the next-smallest c_{ij}, and do the same, taking account of the previous move; continue until all supplies are exhausted and demands met. In our example, the smallest cost is $c_{23} = 1$, so we would send 3 units from supply 2 to fulfill demand 3; we then exhaust supply 2 by sending the other 4 units to demand 1, and complete the initial schedule using supply 1 in the only way possible. And this time, our initial allocation turns out to be the optimal one we have just found, so no introduction of a new route is required.

Recall the assumption that we never find that a partial sum of the supplies adds up to a partial sum of the demands: if two such partial sums are indeed equal, we say the problem is *degenerate*, and we could well find, following the recipe given, that there are *fewer* than $m + n - 1$ equations from which to determine all the u_i and v_j. The method would break down. A subtle trick is available: change each supply from s_i to $s_i' = s_i + \epsilon$, and change demand d_n to $d_n' = d_n + m\epsilon$, where ϵ is tiny. The problem is then non-degenerate, as now no partial sum of the supplies is equal to a partial sum of demands: thus we solve this non-degenerate problem, and put $\epsilon = 0$. Again, an example should help.

Example 4.5 With the same cost matrix as before, let the supplies be 4 and 7, with demands 2, 4 and 5. This is plainly degenerate, so we make the adjustment described, and the northwest corner method leads to $x_{11} = 2$, $x_{12} = 2 + \epsilon$, $x_{22} = 2 - \epsilon$ and $x_{23} = 5 + 2\epsilon$. The initial tableau, after the values of the u's and the v's have been calculated, is thus

$$
\begin{array}{cc}
 & \begin{array}{cccc} d & 2 & 4 & 5+2\epsilon \\ s \quad u_i & v_j & 6 & 5 & 3 \end{array}
\end{array}
$$

$$
\begin{array}{cc}
\begin{array}{cc} 4+\epsilon & 0 \\ 7+\epsilon & 2 \end{array} &
\left(\begin{array}{cc|cc|cc}
6 & 2 & 5 & 2+\epsilon & 8 & 3^* \\
2 & 4^* & 3 & 2-\epsilon & 1 & 5+2\epsilon
\end{array}\right)
\end{array}
$$

Since the values of s and d are automatically catered for in subsequent steps, we can omit specific mention of them. The asterisked value 4 exceeds the corresponding cost 2, so we seek to send a maximal amount θ from supply 2 to demand 1; the other necessary adjustments before fixing the value of θ lead to

$$
\begin{array}{cc}
 & \begin{array}{cccc} u_i & v_j & 6 & 5 & 3 \end{array}
\end{array}
$$

$$
\begin{array}{cc}
\begin{array}{c} 0 \\ 4 \end{array} &
\left(\begin{array}{cc|cc|cc}
6 & 2-\theta & 5 & 2+\epsilon+\theta & 8 & \cdot \\
2 & \theta & 3 & 2-\epsilon-\theta & 1 & 5+2\epsilon
\end{array}\right)
\end{array}
$$

Take $\theta = 2 - \epsilon$, recalculate the $u's$ and the $v's$ to find

$$
\begin{array}{cc}
 & \begin{array}{cccc} u_i & v_j & 6 & 5 & 5 \end{array}
\end{array}
$$

$$
\begin{array}{cc}
\begin{array}{c} 0 \\ 4 \end{array} &
\left(\begin{array}{cc|cc|cc}
6 & \epsilon & 5 & 4 & 8 & 5^* \\
2 & 2-\epsilon & 3 & 1^* & 1 & 5+2\epsilon
\end{array}\right)
\end{array}
$$

Since both asterisked entries do not exceed the corresponding costs, we have found an optimal solution to this perturbed problem: so now set $\epsilon = 0$ to give an optimal solution to the initial problem, i.e. $x_{12} = 4$, $x_{21} = 2$, $x_{23} = 5$.

4.4   Jobs and People

Suppose you have n different jobs to be done, and n people, of diverse suitability, available. You must allocate one job to each person, in the 'best' manner possible. How might you set about this task?

Surprising as it might seem, this problem is identical, in essence, to the transportation problem we have just examined! For each person, give them a score in the range zero to ten, say, according to their suitability for each job: a score of zero means they are totally unsuited, ten means they are ideal. Write a_{ij} as the score of person i to job j, for all values of i and j; make an allocation of the n people to the jobs, add up all the corresponding values of the scores, and the larger this sum is, the better the overall allocation.

So our task is reduced to finding that allocation that maximises this sum: to turn this into the language of the transportation problem, each person is a 'supply' of size 1, each job is a 'demand' of size 1, and if we write $c_{ij} = 10 - a_{ij}$ as a cost, *minimising* the total cost of an allocation is the same as our goal of *maximising* the sum of the a_{ij}. However, it is more natural to stay with the idea of maximising the sum of the values $\{a_{ij}\}$ of the allocations used, and it is not difficult to believe that the only difference in the method of solution is that now we continue to seek a better solution if any asterisked value is *below* the corresponding score $\{a_{ij}\}$.

Plainly this is a degenerate problem! So we adjust it, by having n supplies each of size $1 + \epsilon$, along with $n - 1$ demands of size 1, and one demand of size $1 + n\epsilon$. As ever, an example will prove useful.

Example 4.6 The Head of Department assesses the suitability of four teachers, A, B, C, D to deliver four modules R, S, T, U on a ten-point scale according to the matrix shown, and wishes to allocate one module to each teacher so as to maximise the sum of the associated suitability scores.

$$
\begin{array}{c}
 \\
A \\
B \\
C \\
D
\end{array}
\begin{array}{cccc}
R & S & T & U \\
\left(\begin{array}{cccc}
3 & 7 & 8 & 4 \\
6 & 4 & 5 & 8 \\
7 & 5 & 2 & 7 \\
4 & 6 & 8 & 5
\end{array}\right)
\end{array}
$$

Solution. All the 'supplies' are adjusted to $1 + \epsilon$, the demand for module U is increased to $1 + 4\epsilon$, and we will find a basic (i.e. 7 non-zero entries) solution using the matrix *maximum* method, beginning with $x_{13} = 1$ since $a_{13} = 8$ is a largest entry; the initial tableau, after calculating the $u's$ and $v's$, is

s	u_i	v_j	d	1			1			1			$1+4\epsilon$	
			6			7			8			6		

$1+\epsilon$	0		3	6*	|	7	ϵ	|	8	1	|	4	6*
$1+\epsilon$	-2		6	8*	|	4	9*	|	5	10*	|	8	$1+\epsilon$
$1+\epsilon$	-1		7	1	|	5	8*	|	2	9*	|	7	ϵ
$1+\epsilon$	1		4	5*	|	6	$1-\epsilon$	|	8	7*	|	5	2ϵ

Only at the (4, 3) position is the asterisked value below the corresponding score, so we will put $x_{43} = \theta$, with associated changes $x_{13} = 1 - \theta$, $x_{12} = \epsilon + \theta$ and $x_{42} = 1 - \epsilon - \theta$. We have a new basic solution when $\theta = 1 - \epsilon$. The new tableau, ignoring the values of s and d, is

u_i	v_j	5			7			8			5	
0		3	5*	|	7	1	|	8	ϵ	|	4	5*
-3		6	8*	|	4	10*	|	5	11*	|	8	$1+\epsilon$
-2		7	1	|	5	9*	|	2	10*	|	7	ϵ
0		4	5*	|	6	7*	|	8	$1-\epsilon$	|	5	2ϵ

Now no asterisked value is below the corresponding suitability score, so we have an optimal allocation. Put $\epsilon = 0$: A should take module S, B takes U, C takes R and D takes T, with a total 'suitability score' of 30. (The initial feasible allocation had a score of 29, using the northwest corner method would have given a score with enormous scope for improvement.)

With a little imagination, you can identify many places where this idea could be useful. Some soccer referees will be more suited to certain matches than others; political parties must match the qualities of their candidates to the constituencies where they might stand; charities could aim to make best use of their voluntary helpers when assigning tasks.

4.5   Check Digits

Many organisations use a string of digits to identify their different customers or products. When names are used, it is easy to spot errors—a mistyping that reads 'Pride and Prejduice' is easily spotted and corrected, but you could well not notice an error if you use one of its ISBN numbers such as 1604501480 (its many editions in different countries may have different ISBN numbers). When you make an online payment from your bank account, you are in danger of mistyping an account number and sending money to the wrong person. Not all errors can automatically be picked up, but the use of so-called *check digits* can dramatically reduce the chance of error.

An ISBN 10-digit code is of the form $\{x_1x_2 \ldots x_9x_{10}\}$, where each x_i is a digit in the range zero to nine, except that the last digit, x_{10}, is allowed to take the value X, meaning 10. It is constructed in such a way that

$$10x_1 + 9x_2 + 8x_3 + \cdots + 2x_9 + x_{10} = 0 \quad \text{mod } 11, \qquad (4.1)$$

i.e the sum on the left is divisible by 11. The simplest way to engineer this identity is to use the first nine digits (which gives $10^9 =$ one billion different possibilities!) as the main person or product identifier, then choose x_{10} in the range zero to ten to make the identity hold. In the above example, 1604501480, the sum $10x_1 + 9x_2 + \cdots + 2x_9$ has the value 154, which is already a multiple of 11, which is why $x_{10} = 0$.

To appreciate how useful this is, consider the two most common errors humans are likely to make when writing down a string of ten arbitrary digits: to mistype one number, or to interchange two adjacent numbers. First, if just x_i is incorrectly entered, then the size of the error (positive or negative) is some value m in the range 1–9. In the check-sum, the relevant multiplier is $11 - i$, an integer n in the range 2–10; and since 11 is a prime number, it is not possible for the product mn to be a multiple of 11, so the number typed in fails the check and will be rejected as invalid.

Similarly, suppose two different adjacent numbers x_i and x_{i+1} are inadvertently interchanged when being entered. So instead of the correct expression

$$(11 - i)x_i + (10 - i)x_{i+1}$$

appearing in the check-sum, we find $(11 - i)x_{i+1} + (10 - i)x_i$; and the difference between these two expressions is $x_i - x_{i+1}$ which can never be a multiple of 11.

Not all errors will be picked up: it is quite possible that if two numbers are interchanged, *and* one of them is also entered wrongly, the resulting string happens to satisfy (4.1). But this simple idea, based on properties of prime numbers, will prevent many customers from ordering the wrong book when just using digits to identify it.

4.6 Hierarchies in Large Organisations

The Effects of a Promotions Policy

A firm may wish to know the consequences of their current policies on how many employees are expected to be in each of K grades or salary bands in each of the next few years, or even over the long term. If they can see that the structure would become undesirable at some point, they could alter their policies now to avoid such problems.

Write a_{ij} as the proportion of those in band i who will move to band j in one year, and $A = (a_{ij})$ as the K by K matrix. Plainly we have $a_{ij} \geq 0$ and, for each i, $\sum_{j=1}^{K} a_{ij} \leq 1$. The quantity $1 - \sum_{j=1}^{K} a_{ij}$ is the proportion of those in band i who leave the firm.

Let $n_j(T)$ be the number in band j at time T for $j = 1, 2, \ldots, K$ and $T = 0, 1, \ldots$, so that $\mathbf{n}_T = (n_1(T), n_2(T) \ldots, n_K(T))$ gives the distribution among grades in year T. Finally, let $\mathbf{R}_T$ be the vector giving the number of recruits in year T. We have the recurrence relation

$$\mathbf{n}_{T+1} = \mathbf{n}_T A + \mathbf{R}_{T+1}.$$

Suppose, for simplicity, that all recruitment is to the lowest band, so that, if $\mathbf{e}_1 = (1, 0, \ldots, 0)$, then $\mathbf{R}_T = R_T \mathbf{e}_1$. Thus

$$\mathbf{n}_{T+1} = \mathbf{n}_T A + R_{T+1}\mathbf{e}_1,$$

so that

$$\mathbf{n}_{T+2} = \mathbf{n}_T A^2 + R_{T+1}\mathbf{e}_1 A + R_{T+2}\mathbf{e}_1$$

etc. Hence, for $m = 1, 2, \ldots$,

$$\mathbf{n}_{T+m} = \mathbf{n}_T A^m + \mathbf{e}_1[R_{T+1}A^{m-1} + R_{T+2}A^{m-2} + \cdots + R_{T+m}].$$

If we have a hierarchy with band 1 being the lowest and band K the highest, there is no demotion, and any promotion is just one step up, the only places with non-zero entries in the matrix A are down the main diagonal, and the main superdiagonal: $a_{ii} = p_i$ is the proportion in band i who remain there, $a_{i,i+1} = q_i$ is the proportion who move from band i to band $i+1$ (for $i < K$), so that a proportion $r_i = 1 - p_i - q_i$ of those in band i leave the firm. Thus

$$
A = \begin{array}{c} 1 \\ 2 \\ 3 \\ \vdots \\ K \end{array} \left(\begin{array}{ccccc} p_1 & q_1 & 0 & \cdots & 0 \\ 0 & p_2 & q_2 & \cdots & 0 \\ 0 & 0 & p_3 & \cdots & 0 \\ \vdots & \vdots & \vdots & \cdots & \vdots \\ 0 & 0 & 0 & \cdots & p_K \end{array} \right).
$$

For realism, take $0 < p_i < 1$. Also, to keep the total size of the organisation fixed, as many as leave the firm altogether will be recruited into band 1 for the next year. What happens in the long run?

Note that the model takes no account of how long someone has been in a salary band: so long as the numbers in a band are fairly large, this will not be a serious weakness, as the band will have in it people with a range of service lengths. We are tracking the sizes of bands, not the progress of individuals.

Under our assumptions, $A^m \to 0$ and it seems reasonable to suppose that the numbers recruited each year settle down to some number R, so we expect $\mathbf{n}_{T+m}$ to converge to $\mathbf{n}$, where

$$
\mathbf{n} = R\mathbf{e_1}(I + A + A^2 + \cdots) = R\mathbf{e_1}(I - A)^{-1}.
$$

This shows that the vector giving the long-run distribution is the first row of $(I - A)^{-1}$, multiplied by the annual number recruited. So the key is calculating the matrix $(I - A)^{-1}$.

Example 4.7 Take an organisation of fixed size 10,000 with three salary bands; ten percent of each band leave in any year, ten percent of those in bands one and two are promoted one band. Initially, there are 5,000 in the lowest band, 4,000 in the next, and 1,000 in the top band. Calculate the distributions at the beginnings of each of the next five years (i.e. after recruitment), and the limiting distribution. (Round numbers to the nearest integer.)

Solution. We have $\mathbf{n}(0) = (5000, 4000, 1000)$,

$$
A = \left(\begin{array}{ccc} 0.8 & 0.1 & 0 \\ 0 & 0.8 & 0.1 \\ 0 & 0 & 0.9 \end{array} \right),
$$

and since 10% leave each band, it is plain that $R_T = 1000$ for all times T. Then $\mathbf{n}(1) = (4000, 500 + 3200, 400 + 900) + (1000, 0, 0) = (5000, 3700, 1300)$. Similarly, the rounded figures for subsequent years are
$\mathbf{n}(2) = (5000, 3460, 1540)$, $\mathbf{n}(3) = (5000, 3268, 1732)$,
$\mathbf{n}(4) = (5000, 3114, 1886)$,   and $\mathbf{n}(5) = (5000, 2991, 2009)$.

Also

$$I - A = \begin{pmatrix} 0.2 & -0.1 & 0 \\ 0 & 0.2 & -0.1 \\ 0 & 0 & 0.1 \end{pmatrix},$$

and we seek its inverse. You may already know how to do this, otherwise the Appendix of this chapter shows a simple method starting from scratch. For practise and illustration, here is a calculation via determinants. The determinant of $I - A$ is 0.004, hence its inverse is

$$(I - A)^{-1} = \frac{1}{0.004} \begin{pmatrix} 0.02 & 0.01 & 0.01 \\ 0 & 0.02 & 0.02 \\ 0 & 0 & 0.04 \end{pmatrix} = \begin{pmatrix} 5 & 2.5 & 2.5 \\ 0 & 5 & 5 \\ 0 & 0 & 10 \end{pmatrix}.$$

The long-run distribution is $1000(5, 2.5, 2.5) = (5000, 2500, 2500)$.

Although the limiting distribution depends only on the top row of the matrix $(I - A)^{-1}$, notice that this could change if we change *any* element of A. Suppose we want a long-term structure of $(7000, 2000, 1000)$, and we reckon on an annual turnover of ten percent (not necessarily the same in each band). What matrix A should we use for our promotions policy?

Any A such that the top row of $(I - A)^{-1}$ is $(7, 2, 1)$ will work—but it must be a matrix with all entries non-negative, row sums no more than unity. If we want no demotions (i.e. A is upper triangular), then $(I - A)^{-1}$ will also be upper triangular, so write

$$(I - A)^{-1} = \begin{pmatrix} 7 & 2 & 1 \\ 0 & x & y \\ 0 & 0 & z \end{pmatrix}$$

and we must choose x, y, z. Invert by the determinants method:

$$I - A = \frac{1}{7xz} \begin{pmatrix} xz & -2z & 2y - x \\ 0 & 7z & -7y \\ 0 & 0 & 7x \end{pmatrix}.$$

Taking $x = 2y$ ensures no jump from band one to band three (not necessary, but it reduces the number of unknowns we seek), and then

$$A = \begin{pmatrix} 6/7 & 2/(7x) & 0 \\ 0 & 1 - 1/x & 1/(2z) \\ 0 & 0 & 1 - 1/z \end{pmatrix}.$$

We must have $0 < 2/(7x) < 1/7$, i.e. $x > 2$ for the top row to be sensible. For the second row, we need $1 - 1/x + 1/(2z) < 1$, i.e. $x < 2z$; finally, we need $0 < 1/z < 1$, i.e. $z > 1$. Try $x = 4$, $z = 5$ giving

$$A = \begin{pmatrix} 6/7 & 1/14 & 0 \\ 0 & 0.75 & 0.1 \\ 0 & 0 & 0.8 \end{pmatrix}.$$

It is sensible to check that this does indeed do what we require; this check is left to you. Alternatively, take $x = 8$ and $z = 10$, to give

$$A = \begin{pmatrix} 6/7 & 1/28 & 0 \\ 0 & 7/8 & 1/20 \\ 0 & 0 & 0.9 \end{pmatrix}.$$

Either version of A describes a promotions policy that leads to the desired long-run distribution among the three grades.

Maintaining a Grade Structure

Suppose an organisation has K grades and, in any year, a proportion p_i of those in grade i stay there, while proportion q_i move to the next higher grade. (Plainly, $q_K = 0$, and $p_i + q_i \leq 1$.) Suppose this time that recruitment can be to any grade, not just the lowest. We have

$$A = \begin{array}{c} 1 \\ 2 \\ 3 \\ \vdots \\ K \end{array} \begin{pmatrix} p_1 & q_1 & 0 & \cdots & 0 & 0 \\ 0 & p_2 & q_2 & \cdots & 0 & 0 \\ 0 & 0 & p_3 & \cdots & 0 & 0 \\ \vdots & \vdots & \vdots & \cdots & \vdots & \vdots \\ 0 & 0 & 0 & 0 & \cdots & p_K \end{pmatrix}.$$

If the vector $\mathbf{n} = (n_1, n_2, \ldots, n_K)$ gives the numbers in each band now, then $\mathbf{n_1} = \mathbf{n}A$ gives the numbers before recruitment next year. We want to know whether the distribution $\mathbf{n}$ is *maintainable*, i.e. whether each component of the vector $\mathbf{n_1}$ is no greater than the corresponding component of $\mathbf{n}$ (as then recruitment can regain the distribution $\mathbf{n}$.) Thus $\mathbf{n}$ is maintainable exactly when $\mathbf{n}A \leq \mathbf{n}$.

Suppose $\mathbf{n}$ is maintainable and c is a positive constant. Write $\mathbf{m} = c\mathbf{n}$: thus (using $c > 0$), we see that $\mathbf{m}A = c\mathbf{n}A \leq c\mathbf{n} = \mathbf{m}$: integer considerations apart, it is only the *proportions* in the grades that matter.

Example 4.8

$$A = \begin{pmatrix} 0.8 & 0.1 & 0 & 0 \\ 0 & 0.7 & 0.2 & 0 \\ 0 & 0 & 0.7 & 0.1 \\ 0 & 0 & 0 & 0.8 \end{pmatrix}.$$

Is $\mathbf{n} = (400, 300, 200, 100)$ maintainable? Is $\mathbf{n'} = (500, 300, 180, 60)$?

Solution. Plainly $\mathbf{n}A = (320, 250, 200, 100)$, so this is maintainable: we would recruit 80 in the lowest grade, and 50 into the next lowest.

But $\mathbf{n}'A = (400, 260, 186, 66)$, so this is NOT maintainable—promotion takes too many into the top two grades.

If a structure is not maintainable, it might be *2-step maintainable*: i.e. if we delay recruiting for 2 years, and the sizes of all bands are then no larger than the original numbers, we can recruit to regain the original structure. The condition for $\mathbf{n}$ to be 2-step maintainable is that $\mathbf{n}A^2 \leq \mathbf{n}$, or to be m-step maintainable, it is $\mathbf{n}A^m \leq \mathbf{n}$.

Suppose, for a given A, structure $\mathbf{n}$ is maintainable, i.e. $\mathbf{n}A \leq \mathbf{n}$. As all entries everywhere are non-negative, we can multiply on the right by A to find $\mathbf{n}A^2 \leq \mathbf{n}A$; and hence $\mathbf{n}A^2 \leq \mathbf{n}A \leq \mathbf{n}$, i.e. this structure is also 2-step maintainable.

An obvious induction step leads to the little theorem; 'If a structure is maintainable, it is also *m*-step maintainable, for any $m \geq 1$'.

This raises the question: if a structure is 2-step maintainable, must it be 3-step maintainable? The answer is 'No', even when we have no demotions. For, take A as the 4×4 matrix with $a_{ii} = 0.8$ and $a_{i,i+1} = 0.1$, all other entries being zero. And write $\mathbf{u} = (2916, 1296, 657, 328)$.

Then, by calculation,

$$\mathbf{u}A = (2332.8, 1328.4, 655.2, 328.1),$$

(notice that the top grade is too large, so $\mathbf{u}$ is not maintainable). Calculate

$$\mathbf{u}A^2 = (1866.24, 1296, 657, 328),$$

meaning that $\mathbf{u}A^2 \leq \mathbf{u}$, so $\mathbf{u}$ *is* 2-step maintainable. However,

$$\mathbf{u}A^3 = (1492.992, 1223.424, 655.2, 328.1);$$

and as the final component again exceeds 328, our $\mathbf{u}$ is *not* 3-step maintainable.

4.7 Investing for Profits

Suppose you have funds available to invest in a series of speculative ventures. For simplicity, assume that if you do invest, then either the venture collapses (you lose all your money) or it succeeds (and you double your money). When is it worth investing, and how much should you risk?

Let p be the proportion of times the venture succeeds, and suppose that your criterion is that you wish your funds to increase as fast as possible over the long term. If $p < 0.5$ then, on average, you will lose money, so a policy of investing in those circumstances may pay off some of the time but, in the long run, you will lose.

So suppose $p > 0.5$: even though you succeed more often than you fail, a bold and ambitious approach could easily hit a run of bad luck: if your fortune is F, and you invest all of it every time, on average each investment multiplies your current fortune by $2p$, and $2p > 1$, so on average you achieve exponential growth. However, after

n investments, this average is made up of a tiny chance, p^n, of an enormous fortune, and a much larger chance, $1 - p^n$, of losing everything. It is overwhelmingly likely that you face complete ruin, even though the average outcome appears attractive.

On the other hand, if you are cautious and invest only small amounts, it may take a very long time to make a reasonable profit. So suppose you decide to invest some fraction x of your current fortune: that means that you will invest more next time after a successful venture, less after a loss. What value of x is best?

If your fortune is now F, you invest amount xF, so when you know the outcome, your new fortune is $(1 - x)F$ after a loss, or $(1 - x)F + 2xF = (1 + x)F$ after success. So each time, your fortune changes by the factor $(1 - x)$, or $(1 + x)$, according to the outcome. That means, if you begin with X_0 and have S successes and $n - S$ losses in n ventures, then X_n, your new fortune, comes from

$$X_n = X_0(1 + x)^S(1 - x)^{n-S}.$$

Write $X_n = X_0.r^n$, so that r is the *average* change in your fortune over each of n investments: then

$$r^n = (1 + x)^S(1 - x)^{n-S}.$$

Take logs, divide by n.

$$\log(r) = \frac{S}{n}\log(1 + x) + \frac{n - S}{n}\log(1 - x).$$

When n is large, then S/n, the actual fraction of successful ventures, will be close to p, the long-run proportion of successes. This gives

$$\log(r) \approx p\log(1 + x) + (1 - p)\log(1 - x) = f(x),$$

say. To find the value of x that makes the growth rate r to be as large as possible, differentiate to find $f'(x) = p/(1 + x) - (1 - p)/(1 - x)$. Put the right side equal to zero for a maximum: this gives

$$p(1 - x) = (1 - p)(1 + x),$$

so the solution is $x = 2p - 1$.

And since here $f''(x) = -p/(1 + x)^2 - (1 - p)/(1 - x)^2 < 0$, this is indeed a maximum. We have discovered the *Kelly strategy*, named after John L Kelly who derived it in 1956.

Conclusion: the growth rate of our capital is maximised if we invest the fraction $2p - 1$ *of our fortune.*

As $f(2p - 1) = p\log(2p) + (1 - p)\log(2 - 2p) = \log(2p^p(1 - p)^{1-p})$, the optimum growth rate is then $r = 2p^p(1 - p)^{1-p}$.

Fig. 4.4 Growth rate of fortune against investment size

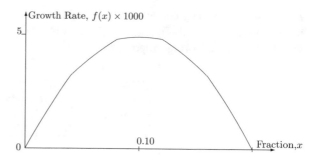

Growth Rate, $f(x) \times 1000$

5

0.10

Fraction, x

0

Example 4.9 Suppose $p = 0.55$. Then since $2p - 1 = 0.1$, we should invest 10% of our current fortune, and the long-term growth rate of our capital is then $r = 2 \times 0.55^{0.55} 0.45^{0.45} = 1.00502$ each time, i.e. about half of one percent.

This may not look wonderful, but it is the best you can do! When $p = 0.55$, the graph of $f(x)$ is shown in Fig. 4.4.

There is a *negative* growth rate if $x > 0.2$, even though the odds are in your favour for each individual investment. So if you greedily risk 25% of your capital on investments with a 55% chance of doubling up, and a 45% chance of total loss, the arithmetic shows that, in the long run, you lose about two-thirds of one percent of your capital each time.

Exercise 4.17 applies these ideas to the context of betting on the spins of a roulette wheel, rather than investing in speculative business ventures—the maths is identical.

4.8 More Worked Examples

Example 4.10 On the front of a car, Goodstone tyres will last for A miles; on the rear, they last for B miles (with $B \geq A$). Suppose you switch front and rear tyres after C miles; what choice of C will make all tyres wear out at the same time, and how many miles will be driven in total at that time?

Solution. After C miles, the front tyres have used a proportion C/A of wear, so when switched to the rear they will be able to travel a further $B(1 - C/A)$ miles, so the total distance travelled will be $C + B(1 - C/A)$ miles. The same argument for the rear tyres show that they will have travelled $C + A(1 - C/B)$ miles; these must be equal, so $B - BC/A = A - AC/B$, i.e.

$$C = AB/(A + B).$$

You should switch tyres after $AB/(A + B)$ miles, and the total distance achieved will be $2AB/(A + B)$ miles.

Example 4.11 A baby cereal to be marketed must contain at least 25% protein, but not more than 70% carbohydrate. It will be formed from a mix of four other cereals, which have the following properties:

Cereal	C_1	C_2	C_3	C_4
% Carbohydrate	80	80	70	60
% Protein	10	15	20	30
Unit cost	1	2	3	4

Set up the task of minimising the cost of a suitable blend as a linear programming problem.

Solution. Suppose we use x_i units of C_i with $x_1 + x_2 + x_3 + x_4 = 1$. To meet the carbohydrate requirement, we need

$$80x_1 + 80x_2 + 70x_3 + 60x_4 \leq 70.$$

For the protein constraint, we must also have

$$10x_1 + 15x_2 + 20x_3 + 30x_4 \geq 25,$$

and we must also have each $x_i \geq 0$. Subject to those constraints, we wish to minimise

$$x_1 + 2x_2 + 3x_3 + 4x_4.$$

Since we have just three independent variables, after eliminating x_1 and simplification, we can rewrite this as: Minimise

$$x_2 + 2x_3 + 3x_4$$

subject to

$$x_3 + 2x_4 \geq 1,$$

$$x_2 + 2x_3 + 4x_4 \geq 3,$$

and $x_2 \geq 0$, $x_3 \geq 0$ and $x_4 \geq 0$.

Example 4.12 Given that the best solution to the previous question is to use just C_1 and C_4 in the ratios $1 : 3$, by how much should the cost of C_2 be reduced so that it would be just as cheap to use a mixture of C_2 and C_4, and what blend would then be used?

Solution. We are given that the optimum mix is $1/4$ of C_1 and $3/4$ of C_4, i.e. take $x_2 = x_3 = 0$ and $x_4 = 3/4$, with the value of the objective function then being $9/4$. If the cost of C_2 were reduced from 2 to c, we would be seeking to minimise $(c - 1)x_2 +$

$2x_3 + 3x_4$, subject to *the same constraints*, of course. Using a mixture with proportion y of C_4 and $(1 - y)$ of C_2, the cost would be $(c - 1)(1 - y) + 3y$, and this would be the same when $(c - 1)(1 - y) + 3y = 9/4$. For the constraints, we need $2y \geq 1$ and $(1 - y) + 4y \geq 3$, i.e. $y \geq 2/3$. Altogether, we have $c = (13/4 - 4y)/(1 - y)$, and $y \geq 2/3$; at $y = 2/3$, we find $c = 7/4$, and c decreases as y increases to its maximum, $13/16$. So use a mix of $1/3$ of C_2 and $2/3$ of C_4, if the cost of C_2 reduces to $7/4$.

Example 4.13 Suppose the roulette wheel in a casino is so biased that it shows Red numbers 60% of the time; a winning bet on Red doubles your money. Find the Kelly strategy that maximises the long-term growth of your fortune, and use the Rule of 72 to estimate how many bets it will take you to double your money. What is the smallest value of x that, if you bet $x\%$ of your fortune each time, you are certain to be ruined?

Solution. In the notation of this chapter, $p = 0.6$ so the Kelly strategy is to bet 20% of your current fortune each time. In that case, the mean growth rate per bet is $r = 2 \times 0.6^{0.6} \times 0.4^{0.4} = 1.02033\ldots$, i.e. about 2% per bet. The Rule of 72 suggests it will take about $72/2 = 36$ bets to double your fortune.

To find the critical point when betting too much leads to losses, solve $0.6 \log(1 + x) + 0.4 \log(1 - x) = 0$, i.e.

$$(1 + x)^{0.6} \times (1 - x)^{0.4} = 1.$$

To solve this, raise both sides to the power five, and rewrite it as $(1 - x)^2 = (1 + x)^{-3}$, giving the iterative scheme

$$x_{n+1} = 1 - (1 + x_n)^{-3/2},$$

which converges to give $x = 0.38939$ or so. If you bet more than (say) 39% of your fortune regularly, you will lose it all in the long run.

4.9 Exercises

4.1 There is steady demand during the year for oil filters. A garage assesses annual demand at 5000 items; the suppliers charge 5 units per item, plus a cost of 50 units per order; the garage carries a holding cost of 2 units per item per year. Sketch the graph of the total cost to the garage, in excess of 25,000 units, as a function of the size Q of each order, and find the Economic Ordering Quantity. Suppose the suppliers give discounts for bulk orders: 0.5% on orders of at least 400, rising to 1% on orders of 500 or more, and 2% on orders of 800 or more. Sketch a new graph, a modification of your previous one, that shows the effects of this discount. What is the optimum now?

4.2 Paul sells pork pies; each pie costs £1.20 to make, the selling price is £3. He has a guaranteed bulk order for 100 pies each day, but he can produce extra to sell to passing customers. Suppose the chance he sells at least K extra pies on any day can be taken as 0.99^K, for $K = 0, 1, 2, 3, ...$ How many extras should he make to maximise his expected profit?

4.3 Use graphs to solve the following problems.

Minimise (a) $3x + y$ (b) $x + 4y$ subject to, in each case,

$2x + y \geq 2,$

$x + 2y \geq 2,$

$6x + y \geq 3,$

and $x \geq 0, y \geq 0.$

4.4 For a problem of the form:

minimise $\alpha x + \beta y$ subject to $x \geq 0, y \geq 0$ and $2x + y \geq 6, x + y \geq 4, x + 2y \geq 5$ (where α and β are constants), sketch a diagram showing the feasible region, and give the co-ordinates of its four vertices.

Solve the problem in the two cases

(a) $\alpha = 5, \beta = 4$: (b) $\alpha = 3, \beta = 7.$

4.5 *Formulate* the following as linear programming problems. (You are NOT asked to solve the problems.)

  (i) Waiters in an ever-open transport cafe work 8-h shifts, one shift per day, and can come on duty at 0700, and every four hours thereafter. The numbers of waiters required in the six periods 0700–1100, ..., 0300–0700 are 10, 14, 8, 10, 3 and 2, respectively. The aim is to employ the smallest possible total number of waiters. Use $\{x_i : 1 \leq i \leq 6\}$ to denote the numbers who start work at 0700, 1100, etc.

 (ii) The Accident and Emergency Department in the hospital is open round the clock. Nurses work for six-hour periods, and begin their shifts at three-hour intervals from midnight. Past experience suggests that the minimum number of nurses required for the eight three-hour periods, starting at midnight, then 3-00 a.m. etc., are 4, 2, 2, 4, 4, 6, 8 and 10. One object is to employ the smallest possible total number of nurses, so long as these constraints are satisfied. Explain your notation carefully.

(iii) The unit costs of food items A, B and C are 25, 30 and 40. Janet wishes to meet her dietary constraints of (at least) 40 units of vitamin R, 20 of vitamin S, 60 of vitamin T and 25 of vitamin U, at minimum cost. The amounts of vitamins R to U, respectively, in unit amount of A are 5, 10, 0, 8; in B the amounts are 0, 6, 15, 3; in C they are 4, 2, 5, 10.

(iv) For $1 \leq i \leq n$, goods of type i each weigh w_i and have value v_i. A container can hold items whose total weight is at most W. Maximise the value of the goods in the container, explaining your notation carefully. (This is known as the *knapsack problem*.)

4.6 Wallpaper rolls are 33 feet long. A particular room needs drops only of lengths 7 feet (full height) and 3 feet (for under the windows).

  (i) List the five sensible ways of cutting any roll to give drops of these lengths.

(ii) The room requires 19 of the longer drops, 8 of the shorter ones. We aim to minimise the total number of rolls to be bought. Formulate this task as a linear programming problem, using $\{x_i : 1 \leq i \leq 5\}$ to denote the respective numbers of rolls that will be cut in the five ways of (i).

(iii) Solve this problem, using only cutting patterns that give at least three drops of length 7 feet. How much wallpaper will be spared at the end?

(iv) Prove that this solution cannot be bettered, even by using other feasible cutting patterns.

4.7 A paper mill produces rolls of standard length, 2.80 m wide. An order is made for 297 rolls that are 110 cm wide, 713 rolls of 91 cm wide, 147 rolls that are 74 cm wide and 301 rolls, each 33 cm wide. Standard rolls can be sliced to meet such orders; the mill seeks to minimise the number of rolls used. Consider five different ways of slicing (measurements in cm):

A: 2 of 110, 1 of 33;   B: 1 of 110, 2 of 74;   C: 3 of 91;
D: 1 of 110, 5 of 33;   E: one each of 110, 91, 74.

Introduce suitable notation, and formulate the mill's task, using just those five cutting patterns. Verify that the order can be met using pattern A 51 times, B twice, C 190 times, D 50 times and E 143 times. With that solution, what proportion of the 436 standard rolls needed is waste?

4.8 Recall that the Standard Minimum Problem is to minimise $c'x$ subject to $Ax \geq b$ and $x \geq 0$. Its *dual*, termed the Standard Maximum Problem, is to choose y to maximise $b'y$ subject to $y'A \leq c'$ and $y \geq 0$.

(a) By rewriting the Standard Maximum Problem as a Standard Minimum Problem, show that the *dual of the dual* is just the original Problem. (Hint: maximising a quantity is the same as minimising the negative of that quantity.)

(b) Write down the duals of the problems found in Q4.3.

4.9 A company has 10 lorry loads of goods in Tilbury, 15 in Harwich, and 15 in Newcastle. It needs to move 24 loads to Liverpool, and 16 to Bristol. The costs, in hundreds of pounds, of moving one load to Liverpool/Bristol, respectively, are: from Tilbury 3 and 2.5; from Harwich 3.2 and 3; from Newcastle 1.8 and 2.6. Advise the company on a strategy that minimises its costs.

4.10 Four cricket umpires, Alan, Brian, Chris and Dickie are to be allocated to matches taking place in Kent, Lancashire, Middlesex and Nottingham, one to each match. The costs of travel and accommodation (in hundreds of pounds) are as shown in the table below.

	K	L	M	N
A	5	6	9	7
B	8	8	6	6
C	5	9	7	8
D	6	5	4	8

Allocate umpires to matches so as to minimise the total cost.

4.11 Suppose that, instead of the decimal system, the world used the *heximal* (or *senary*) system, which has base six. A bank account number of length six would be of the form $x_1, x_2, x_3, x_4, x_5, x_6$ (but where '6' would appear as '10', of course); what would be the corresponding check-sum calculation (use decimals)? Show that Isabella's bank account number of 450232 is indeed valid.

How many different bank account numbers of this length could be issued, if check-sums were used? Interchange the fourth and fifth digits of Isabella's number, AND mistype the sixth, so that the resulting string DOES pass the check-sum.

4.12 We have shown that the check digit formula (4.1) enables us to see whether two adjacent digits have been accidentally interchanged. Show that, in fact, the formula would pick up one accidental interchange of *any* two different digits.

4.13 Employees in a firm are in one of three salary bands $\{A, B, C\}$. Movement between bands takes place annually, at which time 10% of each band leave the firm: of those in band A, 60% stay there, 20% move to band B, 10% to band C; from band B, 80% remain there, 10% move to band C; 90% of those in band C remain—there is no demotion from any grade. Initially, there are 2000 in band A, 1000 in band B, 500 in band C and, at the end of each year, 350 are recruited into band A.

(i) How many will be in each band after the movements and recruitment at the end of two years?

(ii) If E denotes the matrix of movements, verify that $(I - E)^{-1}$ is the matrix

$$\begin{pmatrix} 2.5 & 2.5 & 5 \\ 0 & 5 & 5 \\ 0 & 0 & 10 \end{pmatrix}.$$

(iii) If the pattern described continues indefinitely, what will be the long-term distribution of numbers in each grade?

4.14 In the notation of this chapter, the matrix A below shows the proportions of employees in each of 3 salary bands who stay in that band, or move to the next higher band each year. As many as leave the firm during a year are recruited to band 1 for the next year.

$$A = \begin{pmatrix} 0.7 & 0.2 & 0 \\ 0 & 0.8 & 0.1 \\ 0 & 0 & 0.9 \end{pmatrix}.$$

Suppose that initially the firm has 3,000 employees, all in band 1. Chart the distribution over the next five years (round the number in any band to the nearest whole number). Calculate the long-term distribution of numbers in each band. The firm desires a stable distribution with 1500 in band 1, 1000 in band 2 and

500 in band 3. Construct a realistic matrix that would lead to such an outcome. Assume a total of 3000 employees, and annual recruitment of 300 into band 1.

4.15 With a similar background to the previous question, suppose that the firm wishes to maintain a fixed total number of employees, but may recruit to *any* band to fill vacancies. The matrix giving the proportions who remain, or are promoted in each of the three bands is

$$A = \begin{pmatrix} 0.7 & 0.2 & 0 \\ 0 & 0.6 & 0.2 \\ 0 & 0 & 0.8 \end{pmatrix}.$$

Decide which of the following structures are maintainable, giving reasons for your answers:

(i) (400, 300, 200) (ii) (600, 250, 250) (iii) (800, 400, 420).

Which of these same structures are 2-step maintainable? Say why.

4.16 Suppose the chance that a roulette wheel shows Red is $p = 0.75$, and will pay out winning bets at even money. Describe the policy that maximises the rate at which you expect your fortune to increase. By finding the expected growth rate using the best policy, use the Rule of 72 to estimate how many bets it takes you to double your money.

4.17 Suppose that, as in the text, you will invest a fixed fraction of your current capital in a series of speculative ventures: if a venture fails, you lose the whole amount invested, but if it succeeds you get your investment back, plus a profit of α times the amount invested. Let p denote the proportion of such investments that succeed, with $p > 1/(1 + \alpha)$ so that, on average, you make a profit.

Generalise the text argument to this case, showing that your capital is expected to increase fastest when $x = (p(1 + \alpha) - 1)/\alpha$, and find the corresponding rate at which your capital increases.

In the case when $p = 20\%$ and $\alpha = 5$, estimate how many investments it will take for your initial capital to double.

Appendix Inverting 3 by 3 Upper Triangular Matrices

Suppose the 3 by 3 matrix M is invertible, and 'upper triangular', i.e. it has the form

$$M = \begin{pmatrix} a & b & c \\ 0 & d & e \\ 0 & 0 & f \end{pmatrix}$$

where $adf \neq 0$. Then its inverse M^{-1} also has this format, and it is easy to see that we can put some entries in immediately, and write

$$M^{-1} = \begin{pmatrix} 1/a & x & y \\ 0 & 1/d & z \\ 0 & 0 & 1/f \end{pmatrix},$$

where we seek x, y, z. Since $M.M^{-1} = I$, the $(1, 2)$, $(1, 3)$ and $(2, 3)$ entries lead to

$$ax + b/d = 0 \quad ay + bz + c/f = 0 \quad dz + e/f = 0,$$

so $x = -b/(ad)$, $z = -e/(fd)$, hence from $ay = -bz - c/f$ we find that $y = (be - cd)/(adf)$. Thus

$$M^{-1} = \begin{pmatrix} 1/a & -b/(ad) & (be - cd)/(adf) \\ 0 & 1/d & -e/(df) \\ 0 & 0 & 1/f \end{pmatrix}.$$

You may prefer to use this method with numbers, working out x, y, z from simple simultaneous linear equations, rather than quote an ugly formula.

But, given the matrix A as in Example 4.6, in the matrix $M = I - A$ we have $a = 0.2, b = -0.1, c = 0, d = 0.2, e = -0.1, f = 0.1$; the formula gives its inverse as (reading along rows): $(5, 0.1/0.04, 0.01/0.004)$, then $(0, 5, 0.1/0.02)$ and $(0, 0, 10)$, exactly as before.

Similarly, starting from

$$(I - A)^{-1} = \begin{pmatrix} 7 & 2 & 1 \\ 0 & x & y \\ 0 & 0 & z \end{pmatrix}$$

this formula/method leads to $I - A$ in terms of x, y, z. Immediately we deduce A, and can then choose 'sensible' values of x, y, z as above.

References and Further Reading

Bartholomew D J (1981) Mathematical Methods in Social Sciences (Handbook of Applicable Mathematics) Wiley

Chvátal V (1983) Linear Programming. W H Freeman and Co.

Gilmore P C and Gomory R E (1961) A linear programming approach to the cutting-stock problem. *Operations Research* 9 pages 849-59

Gilmore P C and Gomory R E (1963) A linear programming approach to cutting-stock problems, Part II. *Operations Research* 11 pages 77-91

Haigh J (1992) Further Counterexamples to the monotonicity property of t-step maintainable structures. *Journal of Applied Probability* 29(2) pages 441-7

Kelly J L (1956) A new interpretation of information rate. *Bell System Technical Journal* 35 (July) pages 917-26

Tijms H (2019) Surprises in probability. Seventeen short stories. CRC Press

Trustrum K (1971) Linear Programming. RKP

Vajda S (1978) Mathematics of Manpower Planning. Wiley

Chapter 5
Social Sciences

5.1 Voting Methods

Different countries and organisations use a variety of methods to choose their parliaments, presidents, board of directors or other governing bodies. A sensible voting system must prescribe what the outcome will be, once we know how the votes are cast. We will look at the merits of, and problems associated with, some of these methods. We assume that all voters are able to *rank* the candidates, from first preference down to last (but they are allowed to express no preference between some or all of the candidates). First, we define a few terms.

(a) *Anonymity* means that all voters are treated as equal. If two voters swap their ballot papers, the result is always unchanged.
(b) *Neutrality* means that all candidates are treated as equal. If *every* voter switches the positions of candidates A and B, then the positions of A and B in the outcome would be simply swapped.
(c) *Monotonicity* means that if the only changes to a ballot paper leave a candidate in the same position as before, or place them higher, then it is impossible for that candidate to move down the order.

Most people would agree that these are desirable, maybe even essential, properties that a voting system should possess.

The term *Majority Rule* means that each voter casts one vote; if no candidate gets more than half the votes cast, the outcome is a *tie* (no winner), otherwise the winner is whoever gets more than half the votes cast. It is plain that this system is anonymous, neutral and monotone but, of course, it may produce no winner.

A *quota system* is a system in which there is some number q, the *quota*, such that the winners are exactly those with at least q votes.

Theorem *If there are just two candidates, and the voting system is anonymous, neutral and monotone, then it is a quota system.*

© Springer Nature Switzerland AG 2019
J. Haigh, *Mathematics in Everyday Life*,
https://doi.org/10.1007/978-3-030-33087-3_5

Proof Suppose *n* voters are to choose between candidates *A* and *B* using a system that is anonymous, neutral and monotone, and that if the first *x* voters vote for *A* while the rest vote for *B*, then *A* is elected.

Because the system is anonymous, this means that if *any x* candidates vote for *A*, then *A* is elected. And because it is neutral, we see that if any *x* candidates vote for *B*, then *B* is elected, so if either candidate gets *x* votes, they are elected. Suppose *A* gets *x* votes, and some of the candidates who voted for *B* now switch to *A*: by monotonicity, *A* cannot now be worse off, hence any candidate obtaining at least *x* votes is elected.

Let *q* be the smallest value of *x* such that, if the first *x* voters select *A*, then *A* is elected. Then if *A* (hence also *B*) gets *q* − 1 or fewer votes, they are not elected, but if they get at least *q* votes they are elected—the defining properties of a quota system. The theorem is proved, and the corollary we now state is immediate.

Corollary *In a two-candidate election with an odd number of voters, 'majority rule' is the* only *system that is anonymous, neutral and monotone.*

Effectively, that settles the issue when choosing one of two candidates: if you insist on the method being anonymous, neutral and monotone, then you have to use the 'obvious' method of electing whoever gets most votes. If their total votes are equal, draw lots, or use some other way of giving them equal chances.

But with *n* > 2 candidates, matters are far more complex. In the UK parliamentary elections, when choosing one winner from a list, even though a voter might have well-considered and subtle preferences, only one thing counts—the number of first preference votes. This is known as '*First Past the Post (FPTP)*' or '*Plurality Voting*'. If there are ten candidates, it is quite possible that the one chosen gets only just over ten percent of the votes cast—and the winner might have been the *last* choice of those who voted for somebody else. In the 1992 UK General Election, the outcome in the constituency Inverness, Nairn and Lochinvar found the top four candidates within 3.5% of each other (Table 5.1).

In May 2011, the UK held a referendum on using the *Alternative Vote*, or AV, as a possible change to the current system. Here each voter ranks the candidates: the candidate with *fewest* first place votes is eliminated, and their votes redistributed to their second preference. This is repeated, the remaining candidates move up one place,

Table 5.1 UK General Election 1992: one constituency

Party	Votes	Percentage share
Liberal Democrat	13,258	26.05
Labour	12,800	25.15
Scottish National	12,562	24.68
Conservative	11,517	22.63
Green	766	1.50

Table 5.2 Voting for the
2016 Olympics venue

City	Round 1	Round 2	Round 3
Rio	26	46	66
Madrid	28	29	32
Tokyo	22	20	
Chicago	18		
Total	94	95	98

each time they were originally ranked below the candidate eliminated. Eventually one candidate remains, and we also get a rank order of the candidates.

(Lots are drawn to resolve tied votes, voters need not rank all candidates. The proposal was heavily defeated in the referendum, but this method has been used in other countries, and, indeed, the UK.)

Example 5.1 The 2006 ballot for the leadership of the Liberal Democrats was conducted under AV. The votes cast were

Campbell 23,264; Huhne 16,691; Hughes 12,081.

Thus Hughes was eliminated; his votes went 6,433 to Campbell, 4,937 to Huhne with 711 voters giving no second preference. This gave Campbell 29,697 and Huhne 21,628, so Campbell won.

When the choice of venue for the Olympic games is held, in each round of voting, the electors select one prospective venue, that with fewest votes is eliminated. The voting for the 2016 venue is shown (Table 5.2).

This cannot be AV, as Tokyo's vote dropped from Round 1 to Round 2, so some delegates who gave first preference to Tokyo changed their choice after Chicago was eliminated.

A variant on this system, the *Single Transferable Vote* or STV, can be used to choose more than one winner from a list of candidates. It is used by many Student Unions, and in national elections in Ireland, Australia, Malta and elsewhere. Voters simply mark an order of preference on a single ballot paper, being assured that their higher ranked candidates are not affected by whatever order they choose among those ranked lower. Suppose V votes are cast in total, and N candidates are to be chosen; let $Q = V/(N + 1)$—this is termed the *quota*. Unless Q is an integer, it is plainly impossible for more than N candidates to obtain Q or more first preference votes.

As with AV, count the number of first preference votes for each candidate. Any candidate who scores Q or more is immediately declared elected, and if there are N such candidates, the election is over. At the other extreme, if no candidates obtain the quota, the candidate with fewest first preference votes is eliminated, and their votes are transferred to the next-ranked candidate on each ballot paper.

When any candidate reaches the quota, they may have 'surplus votes' to be transferred to lower ranked candidates. Specifically, if a candidate has $M > Q$ votes, there are $M - Q$ surplus votes to be transferred: to do this in a fair way, each of the M ballot papers is given a weight of $X = (M - Q)/M$, and the next-ranked candidate on each paper is awarded X votes. This may take other candidates over the threshold quota, and they will then be elected, with their surplus votes dealt with in the same way.

The procedure in each round, until N are elected, is: first, check whether any candidate becomes elected, and if that election ends the process; if a candidate is elected, but the process is not ended, transfer any surplus votes; redistribute the votes of the bottom candidate to the next-ranked candidates. Computer programs exist, that take each voter's rank order, and follow these steps, including fractional votes, with complete accuracy. STV reduces to AV if there is just one candidate to be elected.

Example 5.2 Four candidates for two seats: 46 voters, with 13 ranking ABCD, 10 vote BADC, 8 say ACDB, 6 go for CBDA, 5 have DACB and 4 say CABD.

Thus $N = 2$, $V = 46$ and the quota is $Q = 46/3 = 15.33$. First preference votes give 21 to A, 10 to B, 10 to C and 5 to D, so A is elected and has 5.67 surplus votes. Since 13 voters gave the order ABCD, B gets another $13 * 5.67/21 = 3.51$; from the 8 who wrote ACDB, C gets $8 * 5.67/21 = 2.16$. So B now has 13.51, C has 12.16, while D still has only 5 votes. No-one else has the quota, so D is eliminated; all D's voters wrote DACB, but A is already elected, so all 5 votes go to C, who now has 17.16 votes. Thus C gets the second seat.

But suppose that, among the five who ranked D first, x had put B above C, and $5 - x$ had put C above B. Then B would now have $13.51 + x$, C would have $17.16 - x$, so B would be ahead of C if, and only if, $13.51 + x > 17.16 - x$, i.e. $x > 1.83$, in practice $x \geq 2$: B would get the second seat if at least two of D's five voters had ranked B ahead of C.

With *Approval Voting*, each voter can cast one vote for as many candidates as they like—those they 'approve of'. The winner(s) are simply those candidates who attract most votes. Obviously, voting for all candidates has just the same effect as not voting at all, and one objection is that it can encourage negative votes—if you really dislike one candidate, just vote for everybody else. Also, if you would really like A to win, you have no strong feelings about B, but won't vote for C, you have a dilemma: voting for both A and B could make B the winner at the expense of A, voting only for A could let C win in preference to B. Your vote allows no nuances about how strongly you approve of those you don't wish to support.

Two other ways of proceeding carry the names of eighteenth-century French mathematicians/political scientists, Jean-Charles de Borda and Nicolas de Condorcet. In the *Borda Count*, each voter puts the n candidates in rank order from 1 to n, and the winner is selected in one of two equivalent ways: for each candidate, *either* add up the ranks given, and order them according to this total, the lower the better, *or* give a score of $n - j$ points if the candidate is ranked in the jth position, and order them according to the sum of these scores, the higher the better. If two candidates are tied on the same score, some tie-breaking method, such as drawing lots, or favouring the one with the most top rankings, is used.

For any pair of candidates, A and B, each voter is able to say which they prefer, so for each such head-to-head contest, the electorate can pick a winner. If some candidate X beats every other candidate in a head-to-head contest then X is termed a *Condorcet Winner*. If there is a Condorcet Winner, that candidate is elected, otherwise the Condorcet method does not give a winner. If there is no Condorcet Winner, one suggestion is that any candidate who wins *more* pairwise contests than any other candidate should be declared the winner—draw lots if this suggestion produces a tie. (A *Condorcet Loser* is a candidate who loses in every head-to-head contest.)

Plainly, if one candidate is ranked first by more than half the voters, that candidate will be a Condorcet Winner, so if Majority Rule produces a winner, that person is also a Condorcet Winner. However (see Example 5.5), that candidate does not necessarily win under the Borda Count!

Example 5.3 In an electorate of size 23 with three candidates, 5 voters rank them as ABC, 5 as ACB, 7 as BAC and 6 as CAB. So, with first preferences only, A has 10, B 7 and C 6—there is no winner under Majority Rule. But in head-to-heads, A beats B 16-7, A beats C 17-6 while B beats C 12-11, so A is a Condorcet Winner (and C is a Condorcet Loser). The Condorcet method has given a winner when Majority Rule did not.

Example 5.4 With three candidates and 16 voters, 6 voters select the order ABC, 7 choose BCA and 3 choose CAB. Then summing the ranks, A scores 33, B has 28 and C has 35, so the Borda Count gives the overall rank order as BAC. Using Condorcet's method, A beats B 9-7, C beats A 10-6 and B beats C 13-3, so this method leads to a triple tie, with no winner.

Example 5.5 With 61 voters and three candidates, all six possible orderings are used: 10 choose ABC, 16 go for ACB, 13 select BAC, 12 use BCA, 4 offer CAB and 6 choose CBA. Under Condorcet, B beats A 31-30, B beats C 35-26 and A beats C 39-22, leading to the ranking BAC; with Borda, and using the convenient scores of 2, 1, 0 for first, second and third positions, A has 69 points, B has 66 and C has 48. Thus Borda would produce the order ABC.

Note that, if the rules of this election were to use the Borda count, and C had dropped out the day before voting took place, it would be B, not A, who won. Similarly, if all voters moved C to last place, the Borda Count winner changes from A to B—the Borda Count is not *monotone*. The relative ranking of A and B depends on whether or not C is also a candidate. We saw this anomaly in Chap. 3, when looking at Olympic Ice-skating.

The *D'Hondt* method is a form of 'proportional representation'. It is used in elections to the European Parliament, and also in the second stage of elections to the London Assembly. Political Parties offer lists of candidates in their own preferred order, voters cast one vote for their Party of choice. Suppose Party J has $V(J)$ votes, $J = 1, 2, \ldots, n$, and there are to be S seats allocated altogether.

The seats are allocated sequentially, as follows. The first seat goes to the party with most votes. Now suppose Party J currently has $S(J)$ seats; calculate all the

Table 5.3 The D'Hondt method in practice

A	B	C	D	E	Outcome
60,000	51,000	39,000	29,000	16,000	A
30,000	**51,000**	39,000	29,000	16,000	B
30,000	25,500	**39,000**	29,000	16,000	C
30,000	25,500	19,500	29,000	16,000	A
20,000	25,500	19,500	**29,000**	16,000	D
20,000	**25,500**	19,500	14,500	16,000	B
20,000	17,000	19,500	14,500	16,000	A
15,000	17,000	**19,500**	14,500	16,000	C

ratios $V(J)/(1 + S(J))$, and allocate the *next* seat to the party for which this ratio is the highest. Continue until S seats are allocated. As usual, an example clarifies.

Example 5.6 Suppose five parties compete for eight seats; in total, party A gets 60,000 votes, B has 51,000, C has 39,000, D has 29,000 and E has 16,000. We show a table with the ratios $V(J)/(1 + S(J))$ at each stage, the bold figure indicating which party gets that seat (Table 5.3).

Here party A gets three seats, B and C get two each, D gets one and E gets none. A variation on this method due to *Sainte-Laguë*, which tends to favour smaller parties, is described in Exercise 5.5 (ii).

5.2   Voting Dilemmas

We began this chapter by suggesting three properties that any sensible voting system ought to possess. Consideration of systems that are actually used has thrown up other factors. So what general properties should a voting system satisfy? How about:

(1) *Universality* Voters may rank the candidates in any order they wish.
(2) *Monotonicity* No candidate can be disadvantaged by being placed higher on some ballot papers, and not lower on the others.
(3) *Independence of Irrelevant Alternatives* Suppose A, B and C are three candidates. If the system ranks A above B when C is present, it should also rank A above B if C is absent from the list.
(4) *Citizen Sovereignty* There cannot be two candidates A and B with A always preferred to, or tieing with, B, no matter how the votes are cast.
(5) *No Dictatorship* There can be no voter V such that, if V ranks A ahead of B, then the system will rank A ahead of B, no matter how others vote.
(6) *Unanimity* Suppose every voter ranks A above B; then the voting system should rank A above B.

In isolation, each of these looks very reasonable (but we have seen that the Borda Count fails on (3)). Yet economist/political scientist Kenneth Arrow won a Nobel prize in 1972 for his work, leading to

Arrow's First Theorem For an election with more than two candidates, it is impossible for conditions (1)–(5) all to be satisfied.

Arrow's Second Theorem For an election with more than two candidates, it is impossible for conditions (1), (3), (5) and (6) all to be satisfied.

Thus, there cannot be a perfect voting system, if there are more than two candidates! Every system that can be devised has some flaw, we just have to choose between alternatives that are bound to have a deficiency somewhere. Proofs of these Theorems are given in an Appendix to this chapter, following the path described by Hodge and Klima (2005). On a first reading, you may prefer to skip these proofs, but do think about the consequences of this devastating conclusion.

Another result (credited to Allan Gibbard and Mark Satterthwaite) is that, if there are at least three candidates, any of which can win, and the voting system is not simply a dictatorship, then whatever system is used, some voters would be more likely to get their preferred candidate elected if they don't vote according to their preferences—tactical voting!

Example 5.7 Suppose there are three candidates and 100 voters. Their true preferences are: 40 voters would rank ABC, 31 would rank BAC, while 29 would choose CBA.

Under FPTP (or the D'Hondt method), if all vote honestly, A wins. But if the 29 who like C best realise that C can't win, and vote for their second preference, B, then B wins 60-40.

Under AV, if all vote honestly, C gets eliminated in the first round, her 29 votes all go to B, and B wins; but if 3 of those whose real preference is ABC actually vote CAB, then the first round gives 37 to A, 32 to C and 31 to B, who gets eliminated. So when B's 31 votes are transferred to A, A wins 68-32.

With the Borda count, scoring 2, 1, 0 for ranks first, second and third, B wins with 131 against A's total of 111 and C's score of 58; but if more than 20 real ABC voters actually mark their order as ACB, A would win.

Under Condorcet, with honest voting, B beats both A and C, A beats C, so B wins. But if more than 21 of the ABC voters mark their order as ACB, C would beat B and give a triple tie—no winner.

5.3 From the USA

The Constitution of the USA attempts to reflect the general will of the population, but also to ensure that the less populous states are not swamped by the interests of the largest ones. So each state, irrespective of size, elects two Senators, while its number of seats in the House of Representatives is based on its population. Thus, if a state has proportion p of the entire population, and there are to be N House seats in total, that state should have Np House seats—its *standard quota*. But Np will almost certainly not be a whole number! How should this arithmetical fact be overcome?

Definition A system is said to *satisfy quota* if the number of seats to every state differs from Np only by rounding up, or rounding down.

To have a system that satisfies quota is generally thought to be a good thing. One early suggestion was *Hamilton's Method*: find the standard quota for each state, round it *down* to the integer below, and give each state that number to begin with. Count how many seats are left (the surplus seats) and award them one at a time to the states whose standard quotas have the largest decimal parts.

Put simply, the standard quotas with the largest fractional parts are rounded up, the others rounded down. This method plainly satisfies quota. But in 1790, with $N = 105$ the standard quotas for Delaware and Maryland were 1.594 and 8.622, with nine surplus seats. Since $0.622 > 0.594$, Maryland would get a surplus seat (actually, the last to be allocated) before Delaware, giving it nine seats to Delaware's one. George Washington felt this would be unfair to Delaware, as Delaware would get only $1/1.594 \approx 64\%$ its quota, while Maryland got $9/8.622 \approx 104\%$ of its quota. He vetoed the proposal.

An alternative suggested by another Founding Father was *Jefferson's Method*. For a total population of T, write $d = T/N$, call it the *standard divisor*. Note that, if the population of a state is X, then $p = X/T$, so $Np = NX/T = X/d$, hence the standard quota is also calculated as X/d. So first find the standard quotas via X/d: if, by a fluke, all values are integers, we are done, but otherwise, try a *smaller* divisor d', and round *down* the values of X/d'.

If the total number of seats allocated is then N, we have finished—otherwise, try a new divisor. For all X, decreasing d' always increases X/d', so there is a range of values d' that give N seats in total, but they will always allocate the same number to each state. There is no ambiguity.

Example 5.8 In 1822, with total population $T = 8,969,878$ and $N = 213$, New York had a population $X = 1,368,775$. The standard divisor was $d = 42,112$, giving a standard quota of 32.503 to New York. To satisfy quota, New York must have either 32 or 33 seats; but, using Jefferson, $d' = 39,900$ is a suitable divisor, and then $X/d' = 34.305$, so New York would get 34 seats—it doesn't satisfy quota! When a similar problem arose after 1832, suggestions included *Adams's Method*: this is like Jefferson, except that we take $d' > d$, and the values X/d' are rounded *up*, not down. Just as Jefferson's method could give 'extra' seats to larger states, so Adams' method could favour smaller ones (and also not satisfy quota). It was never used. Later came *Webster's Method*: this is also like Jefferson, except that, instead of always rounding up or rounding down, round to the *nearest integer*.

Webster's Method was used in 1842. Exercise 5.7 demonstrates that it is possible for Webster to violate quota. In 1852, it was agreed that Hamilton's Method be used, but also (for safety!) that the total number of seats be such that using Hamilton or Webster gave identical outcomes.

This was not the final word. After the 1880 census, suppose $N = 299$ seats were to be allocated. Then the standard quota for Alabama would be 7.646, for Illinois it would be 18.640, while for Texas the calculation gives 9.640. Looking at the fractional parts of all the standard quotas, Alabama would be placed 20th, and as there were 20 surplus seats, Alabama would qualify for 8 in total, Illinois and Texas would get 18 and 9 respectively.

For a House with $N = 300$ seats, the standard quotas would increase to 7.671, 18.702 and 9.672; so Illinois would be 20th, Texas next—hence Illinois would now get 19, Texas 10, and Alabama only 7! The number of seats available increases, but the number apportioned to Alabama decreases! This is known as the 'Alabama Paradox'. It was resolved by taking $N = 325$, with no such paradox arising.

The books in the References describe other anomalies, including a theorem by Balinski and Young which says that is impossible for an apportionment method always to satisfy quota, *and* be incapable of producing paradoxes.

<u>Hill's Method</u> has been used from 1942. With Webster, a fraction is rounded to the nearest integer, i.e. if K is an integer, and using divisor d' leads to apparent quota x, with $K \leq x < K + 1$, then the cut-off for rounding is at $K + 0.5$, the *arithmetic mean* of K and $K + 1$. With Hill, the cut-off is the *geometric mean*, $\sqrt{K(K + 1)}$.

Example 5.9 Suppose the divisor used leads to apparent quotas of 5.482 for state A and 15.482 for state B. Then Webster would round them to 5 and 15, respectively. But since $\sqrt{5 \times 6} = 5.477$, if we use Hill, then A gets rounded to 6, whereas $\sqrt{15 \times 16} = 15.492$, so B is still rounded to 15.

Why such a fuss over tiny changes in the number of seats in the House that different proposals would produce? One reason is the composition of the Electoral College, that chooses the President: the number of votes cast by a State is the number of its Senators (two), plus its number of House seats, currently giving 538 in total. With an electorate of that size, and States casting their votes en bloc, tiny changes can have large consequences!

5.4 Simpson's Paradox

Two equally effective drugs were tried out on 800 people for side effects. For drug A, 50 out of 200 had side effects—25%, while with drug B, 180 out of 600 had side effects—30%. Drug A came out safer.

The company statistician looked at males and females separately. 150 males had been given drug A, and 30 had side effects—a rate of 20%. 200 males had used B, and 30 of them had side effects—15%. Among *males*, drug B was safer.

By simple subtraction, we deduce that 50 females used A, and 20 had side effects, a rate of 40%; and also that 400 females used B, of whom 150 had side effects, a rate of 37.5%. Drug B was safer among *females* as well!

So B was safer among males, and B was safer among females, but A came out as safer in the whole population—an example of *Simpson's Paradox*. (Dennis Lindley, a prominent UK statistician in the twentieth century, when asked 'If you were at a party, how would you convince people that statistics was more than just "boring numbers"?', immediately cited Simpson's Paradox. See Joyce 2004.)

It seems counter-intuitive that B can be 'better' than A in both subpopulations, but A better than B when the data for these are merged, but this phenomenon is something to watch out for all over the place. In this context, the information to act on is that B is safer, separately, among both males and females, and is thus preferred.

But consider two bowlers on the same side in a two-innings cricket match. Bowler A may have the better analysis in each innings (i.e. a lower ratio of runs to wickets), but B may have the better analysis in the whole match. At a stretch, A could have a better analysis in every one of the ten matches in the season, but B a better analysis over the season as a whole. In this context, we would look at the season as a whole to make the comparison, and agree that B had done better than A.

Depending on context, it may be right to compare A and B among the whole population, or among each of N subpopulations.

One of the best-known examples relates to admissions data to graduate programmes at the University of California, Berkeley, in 1973. Across all programmes, 44% of male applicants were admitted, but only 35% of females. A challenge alleging gender bias failed: within departments, overall admissions rates ranged from only 7% in the most competitive departments (e.g. English) to around 70% in Engineering and Chemistry. But in both those departments, even though far more men than women applied, women had a *higher* success rate; indeed, at the department level, there was generally little difference in success rates. The apparent bias in favour of men was entirely accounted for by the fact that female applications were considerably higher in those departments with lower admissions rates. Representative data are found in Exercise 5.10.

If you express matters algebraically, the appearance of the Paradox is no surprise. For, suppose

$$\frac{a}{b} > \frac{c}{d} \quad \text{and} \quad \frac{e}{f} > \frac{g}{h}.$$

No maths student would dream of 'deducing' that

$$\frac{a+e}{b+f} > \frac{c+g}{d+h}$$

would they?

5.5 False Positives

How useful is screening for breast cancer? Suppose Bessie, a 50-year-old woman with no symptoms, tests positive after a mammograph: how likely is it that she actually has breast cancer?

We plainly need some basic data: take it that around 1% of 50-year-old women with no symptoms actually have breast cancer, and the *sensitivity* of this test is 90%. That means that 90% of women who really do have breast cancer will correctly test positive (and so 10% of them will not). The False Positive rate for this test is about 9%, i.e. of those women *without* breast cancer, 9% will still test positive (so 91% will not). What is your estimate of the chance Bessie has breast cancer? Choose between:

about 9 in 10;   about 8 in 10;   about 1 in 10;   about 1 in 100.

Doctors untrained in the subtleties of statistics frequently get the answer to this question quite wrong. For a valid argument, consider a group of 1000 such women. Since the prevalence rate is about 1%, we take it that this group consists of 10 women with breast cancer, and 990 without. Since the sensitivity of the test is 90%, 9 of those who do suffer will test positive, and 1 will not. But also, 9% of the 990 who do not suffer, i.e. 89 women, will also test positive. So we have $9 + 89 = 98$ women who test positive, but only 9 of them will have the disease. Bessie is one of those 98 women—the chance she is among the nine with cancer is around one in ten.

It is quite erroneous to think that, because the test gives the right answer over 90% of the time, if the test says 'cancer' there is a 90% chance it is correct! On those figures, this test is a disaster: for every woman correctly diagnosed with breast cancer, nine or ten will wrongly be told that they also have the disease, and that will lead to a period of high anxiety until the error is corrected. Increasing the sensitivity from 90 to 100% may look useful, but would make little difference to the downside: now all ten women with the disease will be picked up—but we will still have another 89 wrongly identified. It would be far more helpful to reduce the false positive rate, as then a larger proportion of those picked up will genuinely be sufferers.

In another context, it has been suggested that the use of biometric data (e.g. iris scans) can help detect terrorists (or anyone else seeking to conceal their true identity) at an airport. Identification by machine will not be perfect, but suppose that, when someone uses their correct identity, the machine wrongly raises an alert 1% of the time; and when someone uses a false identity, this is correctly picked up 99.9% of the time. What will be the consequences for, say, the 1,000,000 people who pass through this airport in a given period?

The answer will depend on how many people try to claim false identities. Suppose this is 0.01% of would-be passengers. That means that, among the 1,000,000 people, 100 will be using a false identity, and it is highly likely that all of them will be detected.

But, of the 999,900 honest passengers, 1% of them, or 9,999, will wrongly be stopped as a potential terrorist. The machine picks up $9,999 + 100 = 10,099$ people, but only 100 of them, about 1%, are dishonest. A distressing 99% of all those stopped will, eventually, turn out to be honest.

The single factor that the uninitiated tend to overlook is the base-rate prevalence of the 'condition' under question. The rarer this condition, the more likely it is that a test to identify it wrongly picks out too many suspects.

In general, suppose the intention is to identify people who possess a certain characteristic, X. There are two errors that can be made:

 (i) to assert a person has X, when they do not;
(ii) to assert a person does not have X, when they do.

Without knowing the context, we cannot determine whether error (i) or error (ii) is more serious: and any attempt to reduce the frequency of one type of error will automatically increase the frequency of the other.

For example, in sport, athletes are periodically tested for the use of illegal substances—steroids that build up body strength, drugs that conceal the use of such steroids, 'blood doping' to enhance performance at a particular time. In one context, a quantity known as the T/E ratio is measured—if certain illegal drugs are used, a dramatically high T/E ratio arises. So it would be convenient to use a single cut-off point: a ratio in excess of K is evidence of drug use and punishment follows, ratios below K are acceptable.

But the T/E ratio varies naturally in any individual: a common cold, alcohol consumption, or changing birth control pills, can increase its value. So in the choice of K, a balance must be struck: too high, and drug cheats will prosper, too low, and innocent athletes may have their career unfairly interrupted or terminated. There is no simple answer; see Exercise 5.12.

5.6   Measuring Inequality

It is often claimed that societies would be more socially cohesive, generally happier, if income (or wealth, or property) were more equally distributed. How best might we measure how evenly income (or whatever) is divided? At the heart of several ideas is the *Lorentz Curve*.

To construct this curve (for income), arrange people in increasing order of their income, and suppose that, collectively, the poorest $100x\%$ of the population earn $100y\%$ of the total income. The Lorentz curve is just the graph of y against x, as illustrated as the curve ADB in Fig. 5.1.

The best-known measure of inequality that is derived from this curve is the *Gini Index*, defined as the ratio of the shaded area to the area of the triangle ABC. If all members of the population have the same income, the Lorentz curve coincides with the straight line AB, and the Gini Index is zero (perfect equality); if one member takes the entire income, the Lorentz curve collapses to the two straight lines, AC then CB, and the Gini Index is unity (total inequality): it is about 0.33 in Fig. 5.1.

A fairly common pattern finds a large number of people with low incomes, rather more who are comfortably off, and a small number on high incomes; for convenience, call the bottom 40% the poor, the next 50% the middle, and the top 10% the rich; frequently, that 'middle' tends to earn about half the total income. Figure 5.1 shows that if there is a transfer from the poorest of the middle to the richest of the middle, but not affecting the rich or the poor, the shaded area gets bigger, so the Gini Index

Fig. 5.1 A Lorentz curve

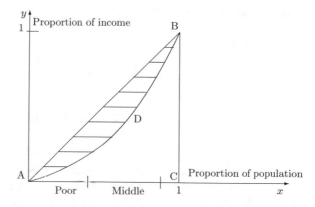

will increase. Social commentators often point to this as a criticism of this Index—it pays too little attention to the extremes of the distribution. But a single number cannot hope to capture all features of inequality.

One quantity that directly addresses the extremes is the *Palma* measure, defined as the ratio of the total income of the top 10% to that of the bottom 40%. With total equality, this ratio will be 0.25, but it is rare to find countries where it is below unity. Broadly, the larger the Gini, the larger the Palma: Denmark has a Gini Index of 0.24, its Palma measure is 0.92, in the UK we find 0.34 and 1.62, while the 2002 figures for Jamaica, then possibly the most unequal country in the world, are 0.66 and 14.67. The Palma measure has a much wider range than the Gini, differences are easier to spot.

The *20–20 ratio* rests on the same idea as the Palma: it is the ratio of the total income of the top 20% to that of the bottom 20%, while a quite different idea is behind the *Hoover Index*—the proportion of total income that should be redistributed to achieve perfect equality. This is also calculated from the Lorenz curve, being the largest vertical distance between the curve ADB and the line AB in Fig. 5.1.

(At some stage in your mathematical career, you are likely to come across the general idea of a *metric space*, i.e. a set of 'points' X and a way of measuring $d(x, y)$, the 'distance' between members of X. In the present context, each 'point' is a continuous curve defined on the interval [0, 1], and we have looked at different ways of defining the distances between two such curves.)

In a population of size N, if the income of the ith person is denoted by x_i and the mean income by $\bar{x}$, the *Theil Index* is given by

$$T = \frac{1}{N} \sum_{i=1}^{N} \frac{x_i}{\bar{x}} \log\left(\frac{x_i}{\bar{x}}\right).$$

If all have the same income, then $x_i = \bar{x}$ for all values of i, and so $T = 0$. At the other extreme, when one person has the entire income, then $T = \log(N)$, so we could normalise this measure by reporting the value of $T/\log(N)$, which will always fall in the range from zero to unity.

To relate this index to the Lorentz curve, consider the ith person: write $p_i = x_i/(N\bar{x})$, the proportion of total income for this person, and $q_i = 1/N$, their proportion of the whole population. Since $x_i/\bar{x} = p_i/q_i$, then

$$T = \sum_{i=1}^{N} \frac{1}{N}\Big(\frac{p_i}{q_i}\Big)\log\Big(\frac{p_i}{q_i}\Big),$$

the format of the Riemann sum approximation to the integral $\int_0^1 g(x)dx$, where g is specified at the points i/N. But notice that p_i is the increase in the cumulative total of wealth attributable to the ith person, while $q_i = 1/N$, so the ratio p_i/q_i will, in the limit for large N, be the *derivative* of the Lorentz curve. This curve is, of course, already a smoothed version of the discrete function that considers each individual separately, so the Theil index is

$$T = \int_0^1 f'(x)\log(f'(x))dx,$$

where $f(x)$ is the Lorentz curve!

Finally, we mention the *Atkinson Index*, defined as $A = 1 - \exp(-T)$, (where T is the Theil Index). Since $T = 0$ with perfect equality of income, so we also have $A = 0$ in that case, and as T rises, so does A, which will thus fall in the range for zero to $(N-1)/N$, effectively unity.

All these measures can be, and are, used in a wide variety of circumstances, not only in the economic field for quantities such as income, wealth or possessions. For example, they offer measures of biological diversity, and differences in educational attainment or quality of life across different countries, regions, or periods of history.

5.7 More Worked Examples

Example 5.10 In an (admittedly peculiar) election, there are three voters, A, B and C, and two candidates X and Y. The outcomes, for three of the possible ways the voters may vote, are as shown (Table 5.4).

(Thus the first row shows that, if A votes for X, and B, C each vote for Y, the winner is X.) Explain why this system is *none of* monotone, neutral or anonymous.

Is there a *dictator*, i.e. someone who's vote always fixes the outcome? Does the person with *fewest* votes always win? Is there *imposed rule*, i.e. the outcome is fixed irrespective of the voting? Show that this is not a quota system.

Solution. It is not monotone: between the first and second row, X gains a vote without losing one, yet changes from a Winner to a Loser.

Table 5.4 Outcomes in an unusual election process

Voter	A	B	C	Outcome
	X	Y	Y	X
	X	X	Y	Y
	Y	Y	X	Y

It is not neutral: in the second and third rows, all voters change their minds, but the outcome is the same—Y wins both times.

It is not anonymous: in both the first and third row, Y gets two votes, but X wins in the first row, Y in the third.

There is no dictator: A's and B's choices lose in row two, C's choice loses in row one.

X has fewest votes in row three, yet Y wins—this is not Minority Rule.

There is no Imposed Rule—it is possible for either to win.

The possible values for the quota are $\{0, 1, 2, 3\}$: if the quota were 0 or 1, all three rows should produce a tie—they don't; if it were 2, Y should win in row one, but doesn't; if it were 3, none of the rows should give a winner, but they all do. It is not a quota system.

Example 5.11 In an election for members of the Senate at Sussex University, under STV rules, there were 52 voters, 8 candidates, and 5 places to be filled. The election outcome was reported as in Table 5.5: what deductions can you make about the voters' ballot papers?

Solution. The Quota is $Q = 52/6 = 8.67$, to two decimal places, so A and E would be elected immediately. The system also elected F, B and D (in that order) We can deduce:

(i) A's surplus was split in the ratios $2 : 1 : 4$, with some rounding. The only way to give these ratios, using integers that sum to at most 11, is that 7 of A's voters made second preferences, with 2 selecting B, 1 for C and 4 for F.

(ii) Since E, with 10 votes, had a surplus of 1.33, and the amount 0.13 was transferred to G, all those 10 voters expressed second preferences, with 5 for D, 1 for G and 4 for H.

(iii) F's surplus was $9.32 - 8.67 = 0.65$, so even if this entire surplus went to C, C would still be bottom, with no further surpluses available. So C must be eliminated, and we can exclude C before distributing the surplus of F.

(iv) We saw that one voter who marked A as first preference also marked C as second; and the only explanation of the 0.33 votes being transferred from C to H is that this voter had H as third preference. Hence just one voter had an order beginning A, C, H.

(v) Two of the four voters whose first preference was H did not express a second preference. So when H was eliminated, it was as though fewer total votes had

Table 5.5 STV in operation

Cand.	First Pref	Surplus of A	Votes	Surplus of E	Votes	C Out	Votes	H Out
A	11	−2.33	8.67		8.67		8.67	
B	6	0.66	6.66		6.66	1	7.66	
C	3	0.33	3.33		3.33	−3.33		
D	6		6	0.65	6.65		6.65	2
E	10		10	−1.33	8.67		8.67	
F	8	1.32	9.32		9.32		9.32	
G	4		4	0.13	4.13	2	6.13	
H	4		4	0.52	4.52	0.33	4.85	−4
Not trans		0.02		0.03	0.05		0.05	2

been cast, so D's score of 8.65 was now enough for election, even if all 'spare' votes went to G.

(vi) Suppose that the first preferences were as above, but only ONE of the 11 who voted for A gave any lower preferences. There would still be 2.33 surplus votes to distribute, but all of them would go to that voter's second preference, giving enormous influence!

Would this voter have wielded more power than can be justified? Effectively, if all others who voted for A gave up their chances to express lower preferences, this one person is acting as representative of A's voters—it is as though all 11 of A's voters had filled in their voting slips in identical fashion!.

5.8   Exercises

5.1 An electorate of size 27 will choose the single winner from four candidates. Each voter *ranks* the candidates, but only four of the 24 possible orders are chosen: 12 voters use the ranking ABCD, 7 use BCDA, 5 choose CDAB and 3 choose DCBA.

   (a) What happens under Majority Rule? Who would win, and what would be the order, under the UK system FPTP?
   (b) How many points would each candidate get under the Borda count? How would this system rank all four candidates?
   (c) These 27 rankings enable you to deduce the choice a voter would make between any two of the four candidates. Give the results of the six possible 'head to head' contests between the four candidates.
   (d) Is there a *Condorcet Winner*? Is there a *Condorcet Loser*?

(e) One possible voting method is *Sequential Pairwise Voting*, SPaV, meaning a sequence of head to heads: e.g. A versus B, loser eliminated, the winner against C; then the winner versus D. The overall winner may well depend on the *order* of these head to heads. Give orders that show that *any* of the candidates could end up as winner.

(f) Explain briefly why SPaV is both monotone and anonymous.

(g) Suppose SPaV is used, in the order (BC), winner versus A, and finally versus D:

   (i) Give the overall ranking.

   (ii) Give the overall ranking, if all 27 voters reversed their preferences as between C and D, nothing else affected.

   (iii) What do your results imply about the neutrality of SPaV?

5.2 (i) Explain briefly why the Alternative Vote (AV) is both anonymous and neutral.

   (ii) AV is to be used to choose one of four candidates, R, S, T and U. Of the 17 voters, 6 choose the order RSTU, 5 choose SRTU, 4 select TUSR, 2 go for UTRS. Conduct the count, and give the result.

   (iii) Suppose the two whose order was UTRS both change their minds, and use URTS (a change that favours R at the expense of T, all others are unaffected). Give the new result.

   (iv) What does this mean about the *monotonicity* of AV?

   (v) Conduct the election for the (original) preferences shown in the previous question under the AV, giving the rank order.

5.3 An electorate of size 46 must choose one winner from the four candidates, A, B, C and D. Each voter ranked all the candidates from best to worst, and 15 voters chose the order ABCD, 13 chose BCAD, 12 selected DBCA and 6 picked CADB.

   (i) Find the rank order using the Borda count.

   (ii) Suppose that, after the votes have been cast, candidate C dies, and it is ruled that C be struck off the ballot papers, and the Borda count be used from the beginning on the remaining three candidates. Give their rank order in these new circumstances.

   (iii) Comparing the relative positions of candidates A and B in your two calculations, describe a deficiency in the Borda count system that this example illustrates.

   (iv) On the original data (i.e. including C), what would be the results of the election under the UK *First Past the Post* (also known as *Plurality*) system, and under *Majority Rule*.

5.4 In the South-East constituency for the European Parliament elections in May 2014, which uses the D'Hondt system, the total votes cast for the parties that won seats were: UKIP 751,439; Conservative 723,571; Labour 342,775; Green 211,876; Lib. Dem. 187,876. Ten seats were allocated by this method: how many seats did each party get, and in what order? If twelve seats had been on offer, who would have got the next two seats?

5.5 In the London Assembly elections of 2012, the four relevant Parties, with their total votes, were: Labour 911,204; Conservative 708,528; Green 189,215; Lib.Dem. 150,447. Eight Labour members and 6 Conservatives were directly elected, and 11 *further* members were chosen, with the *total* number of seats won by each Party following the D'Hondt method, but with a supplementary rule: to qualify for top-up seats, parties must obtain at least 5% of the total votes cast. Since 2,215,008 votes were cast, and UKIP obtained only 100,040 votes overall, they fell short of the qualifying figure of 110,750 (as did some other parties).

   (i) Which parties got these last 11 seats, and in what order?
   (ii) The *Sainte-Laguë* method is a variation on D'Hondt, using the quantity $V(J)/(1 + 2S(J))$ instead, but otherwise proceeding in the same way. What difference would using Sainte-Laguë have made in those elections?

5.6 In the UK 2015 General Election Thurrock constituency, the Conservative candidate won with 16,692 votes, closely followed by Labour (16,156), then UKIP (15,718), with 998 votes cast for other parties. Ignore the votes for the 'Other' parties, and suppose that AV, rather than FPTP had been in use, with voters voting honestly: under what circumstances would the Labour candidate have won the seat?

5.7 The country of Freedomia has ten states, with populations in the proportions 142:237:332:427:522:617:712:807:912:2192 (numbers which sum to 6900). Its parliament has 69 members, the number from each state to 'reflect' that state's population, as in the USA. The standard divisor will thus be 100, giving standard quotas of 1.42, 2.37, etc.

   Give the apportionments that would result using the methods of, in turn, Hamilton, Jefferson, Adams, Webster, Hill. For each, decide whether quota is satisfied.

   Hint: after Hamilton, try the divisors 91.3, 107, 95, 97, respectively.

5.8 A firm has two target groups of customers, the Young and the Old, and two competing sales forces, the Reds and the Blues. Any customer will be approached by one sales force only. Construct *your own* set of numbers for which (a) and (b) hold:

   (a) Reds achieve a higher sales proportion than Blues among both Young and Old separately;
   (b) Overall, Blues achieve a higher proportion of sales than Reds.

   Argue as to which sales force deserves the bigger bonus.

5.9 Construct an example, as mentioned in the text, of two bowlers in cricket where A has a lower runs to wickets ratio in both innings, but B's ratio is better in the whole match.

Table 5.6 Admissions data for Berkeley (1973)

Department	Male applications	Male admissions	Female applications	Female admissions
A	825	512	108	89
B	560	353	25	17
C	325	120	593	202
D	417	138	376	131
E	191	53	393	94
F	373	22	341	24

5.10 The admissions data for the six largest graduate departments at Berkeley in 1973 (see the text) were as shown in Table 5.6.

For each department, compute the proportion of admissions for males and females separately; *without* formal statistical calculations, state whether the admission rates appear to differ in any of these departments.

Compute also the admission rates for men and women across the six departments as a whole, and say whether, on those figures, the rates seem to differ. Comment.

5.11 (i) Among a certain population, 2% suffer from a disease D. There is a test to detect D, but it is not infallible: it gives the correct answer 90% of the time (i.e. whether or not someone is a sufferer, the test has a 90% chance of being correct).

Suppose the population has 1,000 members. Follow through the maths to find the proportion of those who do suffer from the disease, among those the test picks out as sufferers.

(ii) For the same population, change the test reliability so that, of those who actually suffer from D, the test gives the correct result 95% of the time; and among those without the disease, the test gives the correct result 80% of the time. Follow the maths through again.

(iii) Now do it all with symbols: proportion p, with $0 < p < 1$, suffer from D; among sufferers, the test is correct a proportion x of the time, among non-sufferers, it is correct a proportion y of the time.

Among all those picked out by the test, what proportion will actually suffer from the disease?

Hence, under what conditions on x and y will the test be 'useful', in the sense that, among those for whom the test indicates they have D, the proportion who do so exceeds p?

5.12 Suppose that a test to detect the use of illegal performance-enhancing drugs has been developed. It measures a certain quantity, M: for all $x > 0$, among 'clean' athletes, the proportion for whom the value of M exceeds x is $\exp(-x)$, among drug users the proportion is $\exp(-x/4)$. Suppose 10% of athletes tested are drug users: what proportion of those tested will have a value that exceeds x?

Suppose K is taken as the cut-off value, above which a reading is accepted as evidence of drug use. What values of K will ensure that

(i) no more than 5% of clean athletes fail the test;
(ii) at least 50% of drug users are identified?

How many cheats would escape detection if the value in (i) were used? And how many clean athletes would be wrongly accused using the value in (ii)? What value of K minimises the total number of misclassifications, and, for that value, what proportion of those tested will be wrongly classified? (Give all answers to three significant figures.)

5.13 Suppose a Lorentz curve has the format $f(x) = x^\alpha$ over $0 < x < 1$, for some $\alpha \geq 1$. Calculate the corresponding Gini Index, Palma measure, 20–20 ratio, Hoover, Theil and Atkinson Indexes. Evaluate these for $\alpha = 2$.

Appendix

Proofs of Arrow's Impossibility Theorems

Proof Our logic is to prove the Second Theorem, and then deduce the First. We write $A \succ B$ to mean that A is strictly preferred to B, and $A \succeq B$ to mean that A is ranked level with B, or higher. Assume there are $n \geq 3$ candidates, and that the voting system satisfies conditions (1), (3) and (6). We will show that there is some voter, $v*$, who is a dictator, i.e. that (5) cannot hold. To do this, suppose B is a candidate, and every voter ranks B either first or last. Then if the system does *not* rank B as either first or last, there is some candidate, A, with $A \succeq B$, and another candidate C with $B \succeq C$.

Suppose now that every voter makes a simple change, if necessary, to move C above A in their own ranking, making no other changes. Since all voters rank B either top or bottom, this does not affect any voter's preferences between A and B, or B and C; so by property (3), we still have $A \succeq B \succeq C$. However, because every voter ranks C above A, by (6), $C \succ A$ in the voting system—a contradiction. The system *must* rank B either first or last.

So suppose first all voters rank B last. We have just seen that the system must rank B either first or last—and it cannot be first, by (6), so B must be ranked last. Plainly, if all voters now move B from last to first, (6) shows that the system must rank B at the top. Write $\{v_1, v_2, \ldots, v_m\}$ as the list of voters who originally all ranked B last, and now change one at a time in that order to make B first. There is some first voter, v_x say, who's vote change causes B to move from first to last. Call v_x a *pivotal voter*; we show that v_x is a dictator.

For, let A and C be two candidates other than B, and suppose v_x prefers A to C in some list, L, of rankings by all voters. By property (3), the system's choice between A and C must be unaffected if either

 (i) v_x moves B to between A and C, or
 (ii) all of $v_1, v_2, \ldots, v_{x-1}$ move B to the top of their lists, or
 (iii) all of $v_{x+1}, \ldots, v_m$ move B to the bottom of their lists.

So suppose *all* these changes are made giving a new list of rankings L'. In L', $v_1, \ldots, v_{x-1}$ prefer B to A, while $v_x, \ldots, v_m$ prefer A to B. But since v_x is the pivotal voter, as only $v_1, \ldots, v_{x-1}$ rank B first, L' still ranks B last; in particular, L' ranks A above B. But in addition, since B would be ranked top if v_x now moved B to the top from its position of above C but below A, this has no effect on the relative positions of B and C, so L' must prefer B to C. But if $A \succ B$ and $B \succ C$, then $A \succ C$ in L'. But as we know that these changes do not alter the overall choice between A and C, then A is ranked above C in the original list L. Because the pivotal voter v_x prefers A to C, so does the system.

This pivotal voter was chosen by reference to a particular candidate B. So let D be some candidate other than B, and let v_y be the corresponding pivotal voter. Let E be someone other than B or D. We have just seen that whatever choice v_x makes between two candidates other than B—in this case, D and E—the system follows the choice made by v_x. But, as v_y is pivotal for D, v_y also determines the relative ranking of D and E. So v_x and v_y are the same person—there is a single pivotal voter, $v*$ say, who controls the relative rankings of anyone other than B, or D, or E. $v*$ is a dictator.

To deduce the First Theorem, it is enough to show that conditions (2), (3) and (4) imply condition (6); for then, if all of (1) to (5) hold, so must (6), and we have found that (1), (3), (5) and (6) are incompatible.

Given (2), (3) and (4), suppose all voters prefer candidate A to candidate B in some list L of rankings. By (4), we know that there is *some* list L' which would lead to $A \succ B$; modify L' if necessary to produce L'' in which *every* voter prefers A to B. By (2), the system prefers A to B under L''; L and L'' may be quite different, but the relative ranking of A and B is the same for every voter each time so using (3), we see that the system must rank A above B in the original list L. Condition (6) does indeed hold.

References and Further Reading

Arrow K J (1950) A Difficulty in the Concept of Social Welfare. *Journal of Political Economy* 58(4) pages 328–346

Arrow K J (2012) Social Choice and Individual Values. Yale University Press

Bickel P J, Hammel E A and O'Connell W (1975) Sex Bias in Graduate admissions: Data from Berkeley. *Science* 187(4175) 398–404

Cobhham A and Sumner A (2013) Inequality Measures (See *Is it all about the Tails? The Palma measure of Income Inequality*. (Working paper 343, September 2013, Centre for Global Development)

Gibbard A (1973) Manipulation of voting schemes: a general result. *Econometrika* 41(4) pages 587–601

Hodge J K and Klima R E (2005) The Mathematics of Voting and Elections: A Hands-On Approach. American Mathematical Society

Joyce, H (2004) Bayesian thoughts. *Significance* 1(2) pages 73–75

Satterthwaite M A (1975) Strategy-proofness and Arrow's conditions: Existence and Correspondence Theorems for voting procedures and social welfare functions. *Journal of Economic Theory* 10(2) pages 187–217

Taylor A D and Pacelli A M (2008) Mathematics and Politics: Strategy, Voting, Power and Proof. Springer

Chapter 6
TV Game Shows

6.1 Utility

Game shows are a popular TV format. They mix skill and luck in diverse proportions, and some contestants have won enormous prizes. I make no claim that mathematical considerations are a prime factor in determining success. But sometimes mathematics will show you that one particular strategy is more likely to lead to success than another; moreover, an appreciation of non-obvious mathematical ideas can simply add to your enjoyment of the show.

Initially, we will look at shows where contestants have no influence on the size of prize they might win, but when their actions can influence their possible winnings, the concept of *Utility* can guide them into making sensible decisions. Imagine that you have a choice of receiving £1 for certain, or receiving £2 if an ordinary fair coin, when thrown, falls Heads, but zero if it falls Tails. Gambling scruples aside, most people would see those options as about equally attractive—on average, you would get the same amount of money, £1.

But now suppose those amounts are ten thousand times as large: only the very rich would be indifferent, most of us would plump for the definite sum of £10,000 rather than have a 50−50 chances of zero and £20,000. With larger sums, twice as much money is worth *less* than twice as much to us. Our 'Utility' curves have the general shape shown in Fig. 6.1—a larger amount of money is indeed valued higher, initially in a linear fashion, but at a decreasing rate for higher sums.

Given a choice between £10,000 for certain, and £20,000 with some probability p, what values of p make the latter more attractive? Write $U(x)$ as the Utility of amount £x; basing our decision on the idea of Utility suggests that we work out the *average* Utility of the two choices, and plump for the one with the larger value. Taking the gamble leads to $U(20,000)$ a proportion p of the time, and $U(0) = 0$ otherwise, so the average value is the weighted sum of these values, here just $pU(20,000)$. We should take the gamble when p, the chance of success, exceeds $U(20,000)/U(10,000)$.

Example 6.1 Suppose that a contestant in *Who Wants to be a Millionaire?* has successfully answered the £75,000 question. She sees the next question, but is unsure

© Springer Nature Switzerland AG 2019
J. Haigh, *Mathematics in Everyday Life*,
https://doi.org/10.1007/978-3-030-33087-3_6

Fig. 6.1 A typical Utility curve

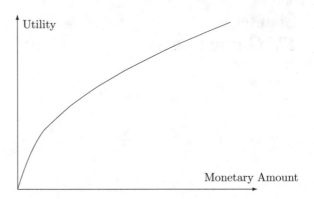

of the answer: if she offers a wrong answer, she will leave with £50,000, but a correct answer takes her to £150,000—and the prospect of more. Should she take the £75,000, or gamble?

Solution. Whatever happens she wins at least £50,000, so she should ask which she prefers, *above* that pleasant prospect: another £25,000 for certain, or (at least) £100,000 with some probability p, zero otherwise. With a choice of four possible answers, even a pure guess will succeed 1/4 of the time: so, if $p > 1/4$, the average amount from the gamble exceeds the sure thing. But our Utility approach tells her to compare the average Utilities, not the monetary amounts, of the different outcomes.

One curve that has the properties described for Fig. 6.1 is $y = K\sqrt{x}$, $K > 0$. The expected utility of taking the £25,000 would be $K\sqrt{25,000} = 158.1K$. If her chance of a correct answer is p, her expected utility in going for the larger prize is $Kp\sqrt{100,000} = 316.2Kp$. 'Utility' indicates that she should take the gamble only if $316.2Kp > 158.1K$, so she should take the gamble only if $p > 50\%$—rather more cautious.

(Of course, a correct answer would give her the chance to increase her winnings further, so the actual utility taking the gamble is *higher* than $K\sqrt{100,000}$; she can justify being a little bolder than this!)

6.2 Monty Hall's Game

The analysis of this game is much better known now than just a few years ago, when it was notorious for causing heated arguments, with holders of PhDs in mathematics making fundamental blunders. 'Monty Hall' was the host for this game, where three closed boxes were on display: one contained a valuable prize, the other two had rubbish. The contestant was invited to select one of the boxes; and when she had done this, Monty (who knew which box the prize was in) would then open one of the other two boxes, showing it contained rubbish, and invite the contestant to stick with their original choice, or swap to the third box. What should she do—or does it make no difference?

We assume the contestant knows this ritual, so she can decide her tactics, to 'Stick' or to 'Swap' in advance. Suppose her decision is 'Stick'. Plainly, there is one chance in three her original choice is correct. The host's action does not affect this, so the 'Stick' strategy wins with probability 1/3.

For the 'Swap' strategy, consider the two times in three that her original choice was one of the rubbish boxes. That leaves two other boxes, one with the prize, and one with rubbish: obviously the host must open the one containing rubbish, so in these circumstances, swapping always leads to the winning box. 'Swapping' gets the prize with probability 2/3.

You win the prize twice as often if you Swap.

All sorts of nonsense have been written in support of other conclusions. For example, some have argued that after Monty has opened a box and shown it to be empty, then as there are just two boxes left, each is equally likely to contain the prize, so whether you Swap or Stick, your winning chance has (magically) increased from 1/3 to 1/2! How Monty's action has changed the probability that your original choice held the prize is not explained.

If your friend is still puzzled when you tell him the correct analysis, given above, change the problem: let there be 100 boxes, 99 with rubbish and one with the prize: after the contestant makes her original choice, Monty solemnly opens 98 boxes, showing rubbish each time. Do they really think each of the two remaining boxes are equally likely to hold the prize?

As in most questions concerning probability, the exact specification of the problem is vital. Here, it is crucial that, *whatever box is chosen*, Monty will *always* open another box. If he only opened a box when the original choice had picked out the prize, swapping would be fatal!

6.3 The Price Is Right

Maths arises at two stages of this show. First, when four contestants are asked to guess the cost, in whole pounds, of some object—a bicycle, a computer, an item of furniture—the winner being whoever gets nearest, *without going over* the actual price. Suppose the guesses of the first three are a, b and c, with $0 < a < b < c$. Then the last contestant should choose one answer from the list $\{1, a+1, b+1, c+1\}$—any other choice would be poor play.

The reason is that if the correct answer has already been chosen, there is nothing the last contestant can do: otherwise, the correct answer is in one of the intervals $[1, a-1]$, $[a+1, b-1]$, $[b+1, c-1]$ or is at least $c+1$. If the correct interval is chosen, the value at the left end of the interval must win—and any other value might not. The key phrase is 'without going over'. The worst answers that could be given are $a-1$, $b-1$ and $c-1$, which can win only if they are exactly right.

The first three contestants do not have this simple selection: indeed, they run the risk that a later player may virtually scupper their chances by guessing £1 more than their offering.

The second place for some maths is at the end of the show, during *The Showcase Showdown*. Three contestants, Adam, Brian and Carla take turns to spin a large wheel, with twenty segments marked $\{5, 10, 15, \ldots, 100\}$; the intention is that, when the wheel is spun, all these numbers are equally likely. The winner is the one with the the largest score, over one spin or the sum of two spins, *provided that* this score does not exceed 100. So each player must decide whether they are satisfied with the first score, or whether they should take a second spin. If two or more of them tie, they have a one-spin play-off, the high score winning. When should each person take a second spin? And what are their respective winning chances?

As often happens in maths, solving an easier problem points the way: so suppose there are just *two* players, and the possible scores are 1, 2, 3, 4, 5, all equally likely. When should the first player, Adam, spin again?

First we must see what the second player, Brian, will do when he knows Adam's score. He will spin again if behind, or if equal to Adam with a score of 1 or 2. (If he equals Adam's score of 3, he will bust 60% of the time if he spins again, so he should go for a one-spin spin-off.)

If Adam's final score is x with $1 \leq x \leq 5$, Brian will win when his first score

 (i) exceeds x, or
 (ii) is x, and he goes on to win, or
 (iii) is less than x, his second spin leads to tie at x, he wins the spin-off, or
 (iv) is less than x, his second spin takes him above x but no more than 5.

The respective chances are

 (i) $(5 - x)/5$;
 (ii) $(1/5).(1/2) = 1/10$ if $x > 2$, or $(1/5).(4/5) = 4/25$ when $x = 1$, or $(1/5).(3/5) = 3/25$ when $x = 2$;
 (iii) $((x - 1)/5).(1/5)(1/2) = (x - 1)/50$;
 (iv) $((x - 1)/5).((5 - x)/5) = (x - 1)(5 - x)/25$.

Summing, we find Brian's winning chances:
when $x = 1$, chance is $4/5 + 4/25 = 24/25 = 48/50$, Adam's chance is 2/50;
when $x = 2$, chance is $3/5 + 3/25 + 1/50 + 3/25 = 43/50$, for Adam 7/50;
when $x = 3$, chance is $2/5 + 1/10 + 2/50 + 4/25 = 35/50$, for Adam 15/50;
when $x = 4$, chance is $1/5 + 1/10 + 3/50 + 3/25 = 24/50$, for Adam 26/50;
and when $x = 5$, chance is $1/10 + 4/50 = 9/50$, for Adam 41/50.

Adam should do these sums. Plainly he will spin again if his first score is 1, but suppose his first score is 2. If he sticks, we have seen that his winning chance is only 7/50; if he spins again, he will bust 2/5 of the time, but otherwise (1/5 of the time each) will reach 3, 4 or 5. His winning chance from spinning is

$$((15/50) + (26/50) + (41/50))/5 = 82/250,$$

much better. Suppose his first score is 3: sticking, his winning chance will be 15/50; spinning, he will bust 3/5 of the time, otherwise he reaches 4 or 5 each with probability 1/5, leading to an overall winning chance of

$$((26/50) + (41/50))/5 = 67/250,$$

which is lower. So Adam should spin again from 1 or 2, stick at 3 or more.

Now we can find their winning chances. If Adam scores 1, he spins, and thus ends up at 2, 3, 4, 5, or bust, all with probability 1/5: from those scores, his winning chances are 7/50, 15/50, 26/50, 41/50 and zero, so the overall contribution is $((7 + 15 + 26 + 41)/50)/25 = 89/1250$.

From 2, he spins and so gets to 3, 4, 5 or bust, leading to a contribution of $((15 + 26 + 41)/50)/25 = 82/1250$.

He sticks at 3, 4 and 5, with respective winning chances of 15/50, 26/50 and 41/50, total contribution $((15+26+41)/50)/5 = 82/250 = 410/1250$. Altogether, his winning chance is $(89+82+410)/1250 = 581/1250$, so Brian's winning chance is 669/1250.

Coe and Butterworth (1995) analysed the real 3-player game with this same approach: they showed that if Carla ties with just one opponent, she should spin with a score of 50 or lower, tieing with both opponents she should spin on 65 or lower; Brian should spin on a tie with Adam at 65 or below, but stop if the tie is at 70 or more; Adam should spin again at 65 or below. Overall, their winning chances are around 31% for Adam, 33% for Brian and 36% for Carla. Maybe closer than you would think?

In the analysis of this game, the fact that two players may end up with exactly the same score complicates matters. But suppose that, instead of a wheel with twenty segments, a 'spin' corresponds to selecting some value completely at random in the interval [0, 1]; each player can take one spin or sum two spins, but a total above unity eliminates the player. Then ties can be ignored. This version, with three (or more) players, was solved by Grosjean (1998). See the appendix.

6.4 Pointless

This show seeks to reward *obscure* knowledge: we can all name some capital city in continental Europe, but what answer might we give if the aim is to name such a city that other people won't think of?

The test used for obscurity of a (correct) answer is to count how many of 100 randomly selected people select that answer within a given time limit. So although answers like Paris or Berlin will be very popular, I believe that Vaduz (Liechtenstein) or Podgorica (Montenegro) would get very few mentions—possibly zero, that much-desired 'Pointless' answer.

Four couples begin the game, three of them are eliminated over a series of rounds based on a wide variety of subjects. In the final round, the remaining pair selects a topic from a list presented to them, and may offer three possible answers: they will win money if *any* of their answers are 'Pointless'. Frequently, they will fail to do so, but do give correct answers chosen only by a small number—two or three—of the 100-strong pool. How unlucky have they been, in the sense that, had a different set of 100 people been asked that question, their answer would have been 'Pointless'.

Let $x > 0$ be the number among the 100-strong pool that gave a particular answer. Then we will take $x/100$ as our estimate of the probability that a randomly selected person will give that answer, so that the probability a person does *not* give that answer would be $1 - x/100$. Hence, with a different set of 100 randomly chosen people, the chance that *none* of them gave that answer is taken as

$$(1 - \frac{x}{100})^{100}.$$

But, when x is small compared to N, we know that

$$(1 - \frac{x}{N})^N \approx \exp(-x).$$

Now $\exp(-1) \approx 0.37$, $\exp(-2) \approx 0.135$ and $\exp(-3) \approx 0.05$; so even if just two people in the pool gave that answer, the chance it would have been pointless with another pool is only about 13.5%. For a single attempt, only if just one person had given their answer might the epithet 'unlucky' be appropriate.

But they offer three answers, so the most frustrating outcome would be if each answer had attracted just one vote. Then, with a different pool of 100 people, each answer has a 37% chance of being pointless, so the chance that *at least one* of them would be pointless is taken as

$$1 - (1 - 0.37)^3,$$

about 75%. And that is as unlucky as you can get in this game.

6.5 Two Tribes

Seven original contestants are whittled down to a single player, who then has the chance to win a cash prize. In each of the first three rounds, the contestants are split into two teams ('tribes') on the basis of a fairly arbitrary criterion, such as do they go on caravan holidays, or whether or not they have the letter 'e' in their name. The teams compete, and one member of the losing team is eliminated. A different criterion leads to new teams in the next round.

When six contestants remain, each tribe has three players, and, all things being equal, every player has the same chance of being eliminated. But with seven or five players, the team numbers are inevitably unequal: is the quiz format fair to all players in those rounds, or does it advantage players in one team over the other? If it is unfair, what changes might make it fairer?

With five or seven players, one team has one more member than the other. Each team, in turn, is asked a series of quiz trivia questions, a correct answer from any team member scores one point, but as soon as the team fails to answer a question correctly, or a one-minute time limit is reached, their turn ends. If the scores are equal, each team selects a 'champion' to go head-to-head to determine the winning team: one member of the losing team is then eliminated (in a fair manner).

Denote the larger team by A and the smaller one by B. Assume all players are equally competent, and let $Pr(B)$ denote the chance that team B wins; plainly $Pr(B) < 0.5$. How big a disadvantage is it to belong to team B?

Look first at the case of five players, with team A having three players and team B just two. Then a given member of B survives if either

(i) team B wins, with chance $Pr(B)$, or
(ii) team B loses, but it is the other player who gets eliminated; the chance here is $(1 - Pr(B))/2$.

Summing these, the chance a given member of B survives is $(1 + Pr(B))/2$.

And a given member of team A survives if

(i) team A wins, with chance is $1 - Pr(B)$, or
(ii) team A loses, but some other player gets eliminated, with chance $2Pr(B)/3$;
(iii) the overall survival chance is $1 - Pr(B)/3$.

The latter survival chance exceeds the former when

$$1 - Pr(B)/3 \geq (1 + Pr(B))/2,$$

i.e. when $Pr(B) \leq 3/5$; a similar calculation for seven players shows that being in team A is advantageous whenever $Pr(B) \leq 4/7$ (see Exercise 6.3). Members of the smaller team are significantly disadvantaged. How best might we overcome this?

We should not be too elaborate here: we *could* set up models of how many points it is reasonable to expect a team of given size to score, and thus estimate some 'handicap' for the larger team. (This general problem is addressed in the last section of this chapter.) But such models will have flaws, and conclusions based on them of questionable value. The critical values of 3/5 and 4/7 are quite similar (60% and 57.1%), but noticeably above 50%, and (by observation) teams score around 7−8 points before they hit the time limit. So (on gut feeling, rather than any specific calculation), we suggest that the smaller team gets a one-point start and is also awarded a win if the scores are tied. This would nudge their winning chance higher: but only data from an experiment would test whether this change evened up the chances of all players.

6.6   The Million Pound Drop

Contestants face a series of questions, each of which has up to four possible answers displayed, but only one is correct. They begin with one million pounds and, with each question, must decide how to split their money among the possible answers. They may select one answer only, or spread it among two or more: any money placed on incorrect answers is lost, and they will win however much they have left after the final question. Two rules: first, *all* the money must be placed somewhere; second, at least one answer must attract *no* money. If they are sure of the answer, they do best to place all their funds on it, but what should they do when they are uncertain between two (or more) answers?

With such large sums available, we again turn to the concept of Utility, using the square root function. Suppose our current fortune is £F with utility $\sqrt{F}$, and we are uncertain as between answers A and B: our intuitive feeling is that they have respective probabilities p and $q = 1 - p$ of being correct, where $0 < p < 1$. Placing £X on A, and £$(F - X)$ on B leads to an expected Utility of

$$p\sqrt{X} + q\sqrt{F - X}.$$

To find the value of X that maximises this, set the derivative equal to zero, i.e.

$$\frac{p}{2\sqrt{X}} - \frac{q}{2\sqrt{F - X}} = 0,$$

or

$$\frac{X}{F - X} = \frac{p^2}{q^2}.$$

In words, the amount you place on any answer should be proportional to the *square* of the chance it is correct. So if you think A is twice as likely as B, put four times as much on A as B, while if you still have one million pounds, and think A three times as likely as B, then split it as £900,000 to £100,000.

The same notion carries through if you wish to spread your money among three answers, say with respective chances 1/2, 1/3 and 1/6: when the dust settles after squaring these values, you should split your funds in the ratios 9 : 4 : 1.

Of course, a different utility function (see Exercise 6.4) might lead to a different decision. A version of this game, where the contestants begin with just £100,000, is now offered.

6.7   Deal or No Deal

This game originated in the Netherlands. In the UK version, 22 sealed boxes each contain a sum of money, ranging from 1p to £250,000, with a mean amount of £25, 712, but a highly skewed distribution—just five amounts exceed the mean, the median amount is below £1000. Each box is allocated to a contestant, one of whom, Chloe, is selected at random to be the main player. Chloe chooses a number of other boxes, their contents are revealed; a 'Banker', Bert, then offers to give Chloe some sum in exchange for her box. If she accepts the offer, the game ends, if she rejects it, more boxes are opened, new offers are made, until eventually either Chloe accepts an offer, or she is left with whatever is in her box. How likely is she to win the top prize?

Consider Chloe's position when just two boxes (including the one holding £250,000) remain. Suppose first that the other box has £50,000 or less. Here Bert is likely to offer at least £100,000, and, rather than risk ending up with a much smaller sum, Chloe will rationally accept (on Utility grounds), forgoing the chance to win the top prize.

But suppose that the other box is also known to contain more than £50,000—it will be either £75,000 or £100,000: with a guarantee of at least £75,000, she may well feel confident enough to take her chance on her own box having the maximum sum.

So to win the top prize, we argue that

  (i) her own box must contain that prize (chance 1/22);
 (ii) the other box, when two are left, must have at least the third prize (conditional chance 2/21)
(iii) she has rejected all previous offers, and
(iv) she rejects the final offer.

Conditions (i) and (ii) alone combine to give a chance 1/231, which is thus an upper bound for the chance of winning the top prize. And this is borne out by the data—in the first 3003 shows, only nine players won £250,000.

It can be instructive to do a thought experiment, and put yourself in the Banker's position, with the following rule: if Chloe accepts your offer, you pay her, and the TV company pays *you* the amount in her box. If she never accepts your offer, the TV company pays her, and you get zero. How would you structure your offers?

At any stage of the game, the different amounts left in the unopened boxes are displayed, and you have some duty to make offers that are 'interesting', otherwise the show will be killed off. It would be perverse to make an offer that exceeded all the possible winning amounts—although this did happen once! If you offer an amount that is *less* than the mean amount in the remaining boxes, then a fundamental result in Probability Theory, known as the Law of Large Numbers, assures you that, in the long run, you will be in profit: this can only happen if your offers are sufficiently tempting to a good proportion of the Chloes who you face and, of course, you need a reserve of capital to guard against a run of bad luck early on.

6.8   The Weakest Link

Here nine initial competitors seek to answer quiz trivia questions. Prize money is built up over a series of rounds, with one competitor eliminated each time; the eventual winner takes all the money. During any round, they face one question each in turn. Suppose Alan has just given the correct answer, leading to possible prize money of X; if the next contestant, Beth, immediately calls 'Bank' before she hears her question, that sum is safely banked, and her question begins a new chain at zero. If Beth does not call 'Bank', and fails to give the correct answer, the money previously accumulated in that chain is lost, but if her answer is correct, the prize money increases to Y, say, and the next contestant Chris must decide whether to bank Y, or seek to build the chain even higher. When should a contestant bank?

When the time limit for that round is almost up, plainly it can be best to bank quickly. But ignoring any pressure of time, the answer is a balance between the amount X that would be banked, and the probability that future questions will be answered correctly. The criterion is to find the banking strategy that maximises the rate at which the prize fund builds up, and the fine detail will depend on the exact sequence of potential prize amounts. Table 6.1 shows the amounts that could be banked according to the number of correct answers in the most common version of the UK game.

If questions are answered independently, with the same chance p of success, and $S(K)$ is the amount banked for a chain of K correct answers, then the mean rate of accumulating money when banking at $S(K)$ can be calculated as

$$\frac{S(K)p^K(1-p)}{1-p^K}.$$

For the array of prize amounts displayed above, this is maximised at $K = 1$ when $p \leq 0.6021$, at $K = 4$ for $0.6022 \leq p \leq 0.7243$, at $K = 6$ if $0.7244 \leq p \leq 0.7953$, at $K = 8$ if $0.7954 \leq p \leq 0.8430$ and at $K = 9$, the maximum available, if $p > 0.8430$. It is never optimal to bank at other prize levels.

(With less precision, this advice is to bank immediately if $p \leq 60\%$, to try to bank after a chain of four or six if $60\% < p < 80\%$, but hope for a chain of eight or nine if $p \geq 80\%$. Exercise 6.6 investigates optimal play for different prize structures.)

Another element in this game is the process of round-by-round player elimination, which is by vote among the remaining players— whoever receives most votes as 'the weakest link' is removed. If Alan and Beth were able secretly to collude and, by a surreptitious exchange of signals, agree to target some particular third player in each

Table 6.1 The prize amounts corresponding to the lengths of chains of correct answers

Answers	1	2	3	4	5	6	7	8	9
Prize (£)	20	50	100	200	300	450	600	800	1000

round, how likely is it that they would end up as the final pair (when they might also have agreed to share the winnings)? As there are nine players, there are 36 possible pairs so, with no collusion, we would put their chance at being the final pair as $1/36$, under 3%.

Perhaps surprisingly, collusion can increase this to over 40%, assuming all the other players simply vote at random for someone other than themselves! To see why this figure might be so large, note that if Alan and Beth survive to be among the final three, they will certainly be able to eject the other player, so look at what would happen if they are two of the final four players, and agree to target Chris. Then, since Chris has at least two votes, the only way Alan might be eliminated is if both Chris and Dinah vote for him (chance $(1/3)^2 = 1/9$) *and* Alan is randomly eliminated on a tie-break with Chris—overall chance $1/18$. The chance Beth is eliminated is the same, so the chance both survive is $8/9$. Similar, but more complex arguments can be constructed for larger groups of players, leading to the answer of some 43.2%.

Collusion aside, the objectively strongest player should never win: at the stage when three players are left, the two weaker players should automatically vote to eliminate the strongest player left, to increase their own chance in the final.

6.9   The Colour of Money

This was a well-hyped but short-lived game. Working in units of £1000, the amounts $\{1, 2, 3, \ldots, 20\}$ were randomly assigned, one to each of twenty boxes identified by their different colours. The contestant was given a target T, in the range $50 \leq T \leq 80$, and won that amount if they could accumulate it by selecting up to ten boxes, one at a time.

When selecting a box, they were aware of the amounts that previously chosen boxes had contained; they made a guess, G, as to the amount in that box. If the box had less than G, they won nothing, if it had G or more, they won that amount G (not necessarily the whole amount in the box). How best to choose the guess G at each stage?

For the first box, there are $G - 1$ boxes with lower amounts, leading to zero being banked, and $21 - G$ boxes with at least G, so the mean amount banked is $G(21 - G)/20$. This achieves its maximum of 5.5 when G is either 10 or 11. On average, the same holds for all ten boxes chosen, so a target of around 50 looks quite achievable, but 70 or more looks rather difficult.

Whatever the target, we claim there must be one or more optimal strategies to reach it! This claim can be proved by the method of *backward induction*. For, suppose we are in the tenth and final round, with eleven boxes left, and some residual target. If that target exceeds the highest amount available, the task is impossible, otherwise the best strategy is obvious.

In the penultimate round, suppose the amounts in the boxes remaining are $x_1 < x_2 < \cdots < x_{12}$, and let the residual target from the last two choices be T_0. We assume that $T_0 \leq x_{11} + x_{12}$, otherwise the target is impossible. We can obviously restrict our choice of G to the 12 amounts that remain; for each choice, we know (by counting!) the chance it succeeds, *and* also the remaining target and boxes for the final round. But as we know the best tactics, and their success chances in that final round, we can now find the success chance for each choice of G, and identify the highest. Thus there is a best strategy (or strategies) for this penultimate round.

Because we know the best tactics for all the situations that might face us with 12 boxes left, parallel arguments hold for the move from 13 boxes to 12, and then from 14 to 13, and so on. The key point is that, at each stage, there are only finitely many choices, and we already know the best tactics and success chances for all the possible outcomes after we have chosen G. There is an optimal strategy from the beginning, but it consists of a huge number of nested 'if' clauses, impracticable to write down. However, we now describe the best strategy for the penultimate round.

Using the notation above, let t be the number of boxes that contain at least T_0. There are two approaches to the penultimate guess if $t > 0$:
(a) try $G = T_0$;   (b) try G as some value less than T_0.
(The strategy for the final guess will be obvious, given the outcome of this initial guess.) To identify the best first step, we *count* how often each option will reach the desired target.

There are 12 choices for the first guess, then 11 (if needed) for the second, giving $12 \times 11 = 132$ choices altogether. We may as well assume all are used, even if the target is met first time. Using strategy (a), there are t choices that will win immediately, and $12 - t$ that give a second chance, so the overall number leading to success is

$$t \times 11 + (12 - t) \times t = t(23 - t). \tag{6.1}$$

There may be several pairs $\{x_i, x_j\}$ with $x_i < x_j$ and $x_i + x_j \geq T_0$, and again we have to examine two cases:
(c) $2x_i < T_0$;   (d) $2x_i \geq T_0$.
In case (c), counting shows that, for any such pair, it is better to try for the higher value, x_j, first, rather than the other way round. There are $13 - j$ boxes that contain at least x_j and if our first choice is one of these, there are $12 - i$ that have at least x_i; on the $(j - 1)$ occasions where the first guess fails, there are still t boxes that get us home in one go, so the overall success frequency is

$$(13 - j)(12 - i) + (j - 1)t. \tag{6.2}$$

In case (d), trying to bank x_i twice is superior to going for both x_i and x_j, and here the success frequency would be

$$(13 - i)(12 - i) + (i - 1)t. \tag{6.3}$$

So the best strategy is to look at the outcomes of (6.1), the possible choices of (i, j) for (6.2), and, if it exists, (6.3); and see which would succeed most often.

Example 6.2 This situation did occur on TV: the last twelve boxes contained the amounts $\{1, 4, 5, 6, 9, 10, 12, 13, 15, 17, 19, 20\}$, and $T_0 = 15$. Then $t = 4$, so the value of (6.1) is $4 \times 19 = 76$. For (6.2), possible choices include $(4, 12)$, $(5, 10)$, $(6, 9)$, corresponding to (i, j) being $(2, 7)$, $(3, 6)$ and $(4, 5)$, with (6.2) having respective values $6 \times 10 + 4 \times 6 = 84$, $7 \times 9 + 4 \times 5 = 83$ and $8 \times 8 + 4 \times 4 = 80$, so the best of these is the first. The pair $(9, 10)$, with $i = 5$, $j = 6$ is a possible candidate for (6.3), but the value is only $8 \times 7 + 4 \times 4 = 72$, so best tactics are to try first to bank 12; if you succeed, go for 4, if you fail go for 15, overall chance $84/132 = 7/11$.

6.10 Who Wants to Be a Millionaire?

A contestant, David, faces a series of quiz questions, the correct answer being one of four possibilities displayed. A correct answer increases the amount won, an incorrect answer ends the game. David has several 'lifelines'—ways of getting assistance— each of which can be used at most once. We assess their value below. Having seen the question and the answers, and even after using some lifelines, David can opt to take the amount won before he saw that question. If he is uncertain of the answer, how should he decide what to do?

The current version has fifteen steps, ranging from £100 to £1,000,000, with one 'safe haven' after five gentle questions at £1000, and another that he can select at a later stage. An incorrect answer ends the game, but he wins the amount in the highest safe haven reached. Earlier, we introduced the notion of *utility* as an aid when deciding whether to guess, or to end the game and take a sure amount.

His prior wealth is crucial in constructing his own utility function. If he is of modest means and, having reached £20,000, he is only about 60% confident of his answer to the £50,000 question, he should probably take that £20,000, rather than risk losing £19,000 of it; but if he is better off, and that loss would not be a disaster, the potential gains if his favoured answer is valid could be decisive.

How valuable are his lifelines? With *Ask the Audience*, each of the 200 or so members of the studio audience votes for their favoured option. If, say, one of the four options gains even 50% of the votes, and no other choice has more than 20%, that is strong evidence that the most popular choice is indeed correct; but a 55–45 split between two choices would be far less reliable. But be warned: once, over 80% of the audience wrongly claimed that alcohol could be consumed in the House of Commons during the State Opening of Parliament (it is only permitted during the Budget speech).

In *Fifty-fifty*, two incorrect alternatives are removed, leaving one wrong answer and the correct answer. This can be decisively helpful—David may know that the remaining wrong answer is incorrect—but it is completely unhelpful if he had already mentally discarded the two choices removed. This option would be very valuable at the first question after a safe haven, if David cannot eliminate any of the four choices: here it can never be wrong to offer some answer, and the opportunity to guess between two choices, rather than four, is very attractive.

With *Phone-a-Friend*, David may select one of several pre-determined, isolated, 'friends', and have some 30 s to obtain help over the phone. He must be brisk, or his time will expire. This lifeline raises an interesting point: if the friend's help is the main factor in getting the correct answer, how much money, in fairness, should David give his friend as a reward? Discuss!

At some versions of the game, one opportunity to *Switch* to a completely new question has been offered. The questions tend to get more difficult as the game progresses, and he will have some reasonable idea of his objective chance of answering questions at that level—maybe 80% at earlier stages, 40% now. Whatever that current figure is, he should be reluctant to Switch if he thinks his chance with the actual question exceeds it, as Switching would be expected to make matters worse.

Another innovation has been *Ask the Host*, where David may seek the help of the question-master, Quentin. Especially when large sums of money are concerned, Quentin is understandably reluctant to completely rule out a possible answer, or state that some answer is definitely correct. Before the quiz, David should do his homework on Quentin, so as to be better able to assess where Quentin has good knowledge. 'Mathematics' has little to offer here.

6.11   Other Shows

In both *The Chase*, and *Eggheads*, we can find that teams of different sizes compete against each other. In pub quizzes, this frequently happens, an incident provoking Dave Percy and Phil Scarf (2008) into suggesting a system of handicapping to take account of this. We will use their model to assess how much additional team members might contribute to an overall score.

Write θ as the chance that a randomly selected team member can give the correct answer to a randomly chosen question. (In practice, θ will vary between the team members, and also with the actual question, but for a simple model we ignore these points.) With a team of size $m > 0$, assuming players act independently, the chance *none* of them can give the correct answer is $(1 - \theta)^m$, so the chance that at least one team member can give the answer is $1 - (1 - \theta)^m$. Table 6.2 shows some numerical values of this probability.

Table 6.2 The chances that a team can give a correct answer according to its size, m and the question's difficulty, θ

θ/m	1	2	3	4	5
0.2	0.2	0.36	0.49	0.59	0.67
0.4	0.4	0.64	0.78	0.87	0.92
0.6	0.6	0.84	0.94	0.97	0.99
0.8	0.8	0.96	0.99	0.998	0.9997

(The original Percy-Scarf application was to a difficult quiz, θ being estimated as below 15%, with teams of up to ten players. With n questions (e.g. $n = 60$), the estimated score of a team of size $m > 0$ is $n(1 - (1 - \theta)^m)$, so adding $n(1 - \theta)^m$ to their actual score might well be a fair handicap.) For the easier questions ($\theta > 0.5$) teams of size three or more are overwhelmingly likely to give the correct answer: the benefit of a large team shows up best with difficult questions (θ small).

The Chase has a resident Chaser (a quiz trivia expert) who faces four contestants. Each contestant, in turn, answers a series of quiz questions, a correct answer gaining a nominal £1000. Perhaps Maria earns £4000 here. The default position is that she and the Chaser will then face a series of identical quiz questions: she begins five steps from Home, the Chaser starts two steps behind her, correct answers move either one step closer to Home. If Maria reaches Home before the Chaser catches up with her, the £4000 goes forward to the last stage (the 'Final Chase'), otherwise that money is lost and she is eliminated.

The Chaser will offer Maria two alternatives: if she is content with some lesser sum, say £1200, she may start one step closer to Home, or does she wish to try for a much larger sum, say £30,000, but starting one step further from Home. How should she respond?

All the contestants who reach Home ahead of the Chaser will form a team, who will win (and share equally) the total amount they carried through, provided they can answer *more* questions than the Chaser in that Final Chase. A larger team might markedly increase the number of correct answers they give: a reasonable estimate of θ, the chance that a random contestant knows the answer is in the range from 50% to 75%. Maria's decision will take account of how much money previous contestants have carried through.

For example, if some £40,000 is already in the pot for the Final Chase, it makes far more sense for Maria to be unambitious, and go for the £1200 offer, rather than stay with the default £4000, as having £41,200, as opposed to £44,000 hardly changes the Utility, but increases the chance of having a larger team in the Final Chase. Maria should also think seriously about the offer of £30,000, as having £70,000 rather than £40,000 might be quite exciting. It would be poor play for Maria to stick at the £4000 default option.

But if the prize fund is very low when Maria has to decide, a reasonable case can be made for any of the three alternatives.

In both *Shafted* and *Golden Balls*, the game ends with two players and a potential prize fund. After some discussion about *trust*, they must each, simultaneously, state whether they wish to split that fund equally ('Share'), or take it all themselves ('Steal'). If both opt to Share, the fund is indeed split equally; if one is willing to Share, but the other will Steal, the Stealer takes the lot; if both would Steal, neither gets anything.

You may well recognise this as a variant on the much-discussed 'Prisoners' Dilemma'. Here each may think along the following lines: 'If my opponent chooses Steal, I will get nothing whatever I do. If he chooses Share, I will get twice as much if I Steal. So whatever he does, I am at least as well off if I Steal, and I might be better off. So I will Steal'. And if this scene plays out, both using that chain of thought, both will opt to Steal, and both will lose out.

6.12 Exercises

6.1 Generalise the 'Three Boxes' problem to the case when there are N boxes ($N \geq$ 3), one containing the prize, and $N - 1$ being empty. After you have chosen one of these N boxes at random, the host will open r boxes ($1 \leq r \leq N - 2$), showing them to be empty, and invite you to stick with your original choice, or to swap to one of the remaining $N - r - 1$ boxes, chosen at random.

For the two tactics, 'Stick' or 'Swap', give the respective chances of ending up with the prize, in terms of N and r. Make your recommendation.

Finally, analyse the case when m of the N boxes contain a token leading to the valuable prize. The same framework applies, and we have $m \geq 1, 1 \leq r \leq N - m - 1$.

6.2 Give a complete analysis of the Showcase Showdown for two players when there are just four values, $\{1, 2, 3, 4\}$ on the wheel.

6.3 Confirm the claim made in the text that in the game 'Two Tribes' with teams of size four and three, all players equally proficient, being in the smaller team makes it more likely that you will be eliminated in that round, if the chance that team wins is less than $4/7$.

6.4 Suppose your utility function for the amount $£X$ is $X^{1/3}$. In the Million Pound Drop, how should you split your current fortune so as to maximise your expected utility, if you are torn between two answers, but think one twice as likely to be correct as the other?

6.5 A number of (fun) versions of Deal or No Deal can be found on the internet. Find one, play it, and report what happened in one play of the game, playing through to the end, justifying your decisions, to Deal or Not to Deal, in each round.

6.6 In 'The Weakest Link', and using the notation given in this chapter, suppose you use the criterion of banking when

$$R(p, K) = S(K)p^K / (1 + p + p^2 + \cdots + p^{K-1})$$

is a maximum (i.e. ignoring time factors).

(a) If the prize levels increase as $S(K) = K$, show that you should bank immediately, whatever the value of p.

(b) Suppose $S(K) = \alpha^K$, where $\alpha > 2$ and $K = 1, 2, \ldots, K_0$. By considering when $R(p, K + 1)$ exceeds $R(p, K)$:

  (i) show that, if $p \leq 1/\alpha$, it is optimal to bank at $K = 1$;
  (ii) show that, if $p = \beta_K$ is the solution of $\alpha p - 1 = (\alpha - 1)p^{K+1}$ when $p > 1/\alpha$, then $\beta_1 > \beta_2 > \ldots > \beta_{K_0} > 1/\alpha$, and banking at $K + 1$ is better than banking at K if, and only if, $p > \beta_K$;
  (iii) deduce that, if $p > 1/\alpha$ and $p > \beta_1$, then it is optimal to bank at $K = K_0$;
  (iv) explain why, if $\beta_K > p > \beta_{K+1}$, banking at $K + 1$ is the *worst* decision;
  (v) deduce the optimal banking decision for all p with $0 < p < 1$.

6.7 Suppose the last twelve boxes in The Colour of Money contain the values $\{1, 2, 4, 5, 8, 9, 10, 11, 12, 13, 16, 20\}$. Find the optimum tactics in the cases where the residual target is: (i) 10; (ii) 12; (iii) 16.

6.8 In the game 'Who wants to be a Millionaire', explain what you would do in each of the following positions, and why.

(a) You have used all lifelines except 50–50, could take £5,000, the next question is worth £10,000, and all four answers look plausible.
(b) The same, except that you have £75,000 and the question is worth £150,000.
(c) The same, except that you have £500,000 and the question is worth £1,000,000.
(d) The same as (c), except that your remaining lifeline is phone-a-friend, and your friend says she is 75% certain of the answer she gives.

6.9 In the game 'Who wants to be a Millionaire', rank the following breakdowns of the voting in 'Ask the Audience' in order of helpfulness, and justify your answer. (Figures are percentages.)
(i) 60, 15, 15, 10: (ii) 55, 40, 5, 0: (iii) 50, 20, 15, 15: (iv) 50, 30, 10, 10.

Appendix

Following Grosjean (1998), we analyse the three-person 'Spinning' game, with a wheel giving values in the continuous range $[0, 1]$; each player may take one spin, or two spins, but is eliminated with a score of zero if the second spin takes the total above unity.

Suppose an early player finishes with a total of x, with $0 < x < 1$. Any subsequent player beats this, *either* by scoring more than x on the first go (probability $1 - x$ of course), *or* by scoring some $y < x$ on the first go (probability x), and then scoring z, with $x - y < z < 1 - y$, on the second go (Fig. 6.2).

But notice that, if this second go is needed, the length of the 'winning interval' is $1 - x$, whatever the value of y. Thus, overall, a subsequent player will exceed x with probability $(1 - x) + x(1 - x) = 1 - x^2$. And hence a score of x has chance x^2 of beating any one subsequent player.

We now work backwards. Suppose the best outcome by the first two is x. The argument above shows that the chance Carla wins is $1 - x^2$.

What will x be? Suppose Brian has a score of y, already bigger than Adam's score. Should he spin again? If he does not, his winning chance is plainly y^2, so see what happens if he does take a second spin.

Fig. 6.2 Early player scores x, later player scores $y < x$

With probability y, his next spin scores more than $1 - y$ and he is eliminated; otherwise, his next spin is z ($z < 1 - y$), his total is $y + z$, making his winning chance $(y + z)^2$. If he spins, his winning chance is

$$\int_0^{1-y} (y + z)^2 dz = [(y + z)^3/3]_0^{1-y} = (1 - y^3)/3.$$

So he should spin whenever $(1 - y^3)/3 > y^2$, i.e. whenever $y < 0.5321 = \alpha$, say.

So Brian spins whenever he is behind Adam, or when he is ahead of Adam, but his score is less than α.

Adam can work this out, so what action should he take, with an initial score of z? If $z < \alpha$, he must spin again—he would do so to maximise the chance of beating one opponent, let alone two. So assume $z \geq \alpha$. Should he spin?

If he does not spin, he wins only if both Brian and Carla fail to exceed z. For each of them, the chance they don't beat z is z^2: the overall chance they *both* fail to beat z is z^4. Adam's winning chance if he does *not* spin is z^4.

Suppose Adam spins from his score of z. To win, he must score y, with $y < 1 - z$, and then beat both Brian and Carla from his score of $z + y$. Since we have assumed that $z \geq \alpha$, plainly $z + y \geq \alpha$, so his winning chance would be $(z + y)^4$. Overall, his chance is

$$\int_0^{1-z} (y + z)^4 dy = (1 - z^5)/5.$$

He should spin again if $(1 - z^5)/5 > z^4$, which occurs if $z < \beta = 0.6487$.

In summary: Adam should spin if he scores less than $\beta = 0.6487$; Brian should spin if his first score is less than Adam's, or if he beats Adam, but his score is less than $\alpha = 0.5321$; and Carla should spin if her first score has not already won.

We now find their respective winning chances. Calculation is a little complex, but a good computer will spit out as many random numbers, i.e. scattered uniformly over the range $(0, 1)$, as we like, so Grosjean *simulated* this game 20 million times to get reliable answers. The respective winning chances came out as close to 30.5%, 33% and 36.5%, very close to the values found by Coe and Butterworth in the real game.

For the n-player game, using the same argument, Adam should spin again if his first spin gives less than γ, the positive root of

$$x^{2n-1} + (2n - 1)x^{2n-2} - 1 = 0.$$

References and Further Reading

Coe P R and Butterworth W (1995) Optimal Stopping in "The Showcase Showdown". *The American Statistician* 49(3) pages 271–5.

Grosjean J H (1998) Beating the Showcase Showdown. *Chance* 11(1) pages 14–19

Haigh J (2003) The weakest link. *The Statistician* 53(2) pages 219–26.

Haigh J (2014) Pointless: The maths of TV Game Shows. *Plus Magazine*

Haigh J (2015) Making Two Tribes Fairer. *Plus Magazine*

Percy D F and Scarf P A (2008) On the development of decision rules for bar quiz handicapping. *Journal of the Operations Research Society* 59(10) pages 1406–14.

Thomas L C (2003) The best banking strategy when playing the Weakest Link. *Journal of the Operational Research Society* 54(7) pages 747–50

Wolstenholme L and Haigh J (2006) Deal or no deal? *Significance* 3(4) pages 191–2

Chapter 7
Gambling

7.1 Introduction

The organisers of commercial gambling opportunities—casinos, bingo halls, lotteries, sports betting, etc.—do so in the expectation of making a profit. That they continue to flourish shows this expectation is fulfilled. In games of pure chance, the prize levels are set to pay out less, and often considerably less, than the entrance fees paid by the public. But sometimes, there is enough opportunity to show skill, meaning that some participants can legitimately expect the odds to be in their favour; in poker, a top player may be sure that, in the long run, her winnings will exceed the entrance fees, in horse racing, a very knowledgeable punter will identify times when the terms offered by the bookmakers underestimate the true winning chances. But most of the time, the more you gamble, the more you lose.

Suppose all six outcomes when you roll a die are taken as equally likely. Then the *probability* of each outcome is $1/6$; equivalently, we can say that the *odds against* any of them are 5–1. But in gambling, the term 'odds' is more frequently used to mean something quite different, describing the amount a bookie or casino will pay if your bet is successful. For example, a bookie may offer 'odds' of 8–1 that Manchester United will beat Arsenal $2-1$; that means that if your bet is successful, your stake, plus eight times your stake, will be returned to you. Here there is no automatic conversion to give the probability of that particular score. Rather than 'odds', a much better term is 'payout price', but you must get used to seeing the word 'odds' in that context.

In a soccer match, the payout prices for a Home win, Away win or Draw may be expressed as 5–6, 4–1 and 12–5, respectively. If your bet of six units on a Home win is successful, you will receive back 11 units altogether—the six units you bet, plus five more. A successful bet of five units on a Draw would lead to a return of $17 = 5 + 12$ units. In all cases, 'odds' can be expressed in the form α to 1: here $\alpha = 5/6$ for a Home win, or $\alpha = 12/5 = 2.4$ for a Draw. For the Home win, with $\alpha < 1$, this would be termed 'odds on'; when $\alpha > 1$, the term is 'odds against'. With

© Springer Nature Switzerland AG 2019
J. Haigh, *Mathematics in Everyday Life*,
https://doi.org/10.1007/978-3-030-33087-3_7

$\alpha = 1$, which applies to many popular bets in roulette or coin-tossing, the term is 'evens', or 'even money'.

In most of the applications discussed in this chapter, the probability of an event of interest will indeed arise in the format m/N, where there are N possible outcomes altogether, taken as equally likely by symmetry, and the event of interest corresponds to exactly m of them. But sometimes, we shall want to use the term 'probability' in a wider context: informally, in this case the simplest interpretation of the word is the long-run relative frequency—if we were able to carry out an experiment many times under identical conditions, the proportion of those experiments in which our event occurred will settle down to what we term its probability.

7.2 Lotteries

Commercial lotteries are based on the principle that all the 'tickets' sold have the same chance of success. This is achieved in diverse ways, but to analyse their workings, the main tool is simple counting. A large proportion of the adult UK population buys lottery tickets, the most popular being the National Lottery and Euromillions, both of which offer the chance to win millions of pounds. The fundamental mathematical fact in the analysis of these games is that the number of ways of selecting r objects from a set of n objects (without regard to their order) is

$$\binom{n}{r} = \frac{n!}{r!(n-r)!}. \tag{7.1}$$

For its first 21 years, the UK National Lottery followed the format termed 6/49, as six (winning) numbers would be chosen from the list $\{1, 2, 3, \cdots, 49\}$, all possible choices being equally likely. Formula (7.1) shows that this gave $N_1 = 13,983,816$ different choices. Gamblers also chose six of these numbers, and would win a prize if at least three of their selections were among the winning numbers. If they selected exactly r of the six winning numbers, then they would have also selected $6 - r$ of the other 43 numbers; overall, there are just

$$\binom{6}{r}.\binom{43}{6-r} \tag{7.2}$$

ways of doing this.

In late 2015, the format changed to 6/59; Formula (7.1) leads to $N_2 = 45,057,474$ different choices, and now a prize is won if at least two selections are winning numbers. Plainly, Equation (7.2) is modified by replacing '43' by '53' to give the information for the new format. Table 7.1 shows the different numbers of ways of matching r winning numbers.

In the 6/49 format, there are 260,624 ways of matching at least three winning numbers, so the chance of winning some prize with a single ticket was $260,624/N_1$,

Table 7.1 The numbers of ways of matching r winning numbers

No. of winning numbers	No. of Ways (6/49)	No. of Ways (6/59)
6	1	1
5	258	318
4	13,545	20,670
3	246,820	468,520
2	1,851,150	4,392,375
1	5,775,588	17,218,110
0	6,096,454	22,957,480

about 1 in 54, while the chance of winning a share of the jackpot was just $1/N_1$, about 7×10^{-8}. But, under 6/59, as there are 4,881,884 ways to match at least two winning numbers, the chance of winning *something* has increased to 4, 881, 884/N_2, just under one in nine, while the chance of a share of the jackpot is $1/N_2$, less than one- third as much as before. The 'prize' for matching two numbers now is a free 'Lucky Dip' ticket to the next Lottery; the chance of a *cash* prize, by matching three or more winning numbers, has dropped to 1 in 92.

Both versions follow a common pattern in distributing the prize money: a certain proportion of the total generated by sales is returned as prizes, and the amount of the bottom cash prize is fixed. The remaining prize money is allocated in fixed proportions to the other prize-winning levels, and, at each such level, the prize amount is split equally among all winning tickets. There is thus no guarantee that matching x winning numbers will yield a larger prize than matching $x - 1$. Indeed, it has happened in the UK Lottery that the jackpot prize has been *less* than the 'second' prize, won by matching five numbers and the Bonus Ball. See Exercise 7.1.

An intriguing question is how many tickets must be bought in order to guarantee winning at least one prize. In the 6/49 format, needing to match at least three winning numbers, it has been shown that a particular choice of 163 tickets carries this guarantee and, at the time of writing this is the record-holder (see, e.g., the 'wheeling challenge' at the website https://lottery.merseyworld.com).

For the 6/59 format, it is possible to guarantee to match at least two winning numbers with 30 tickets. For, since there are six winning numbers, at least one of the five lists $\{1, 2, \ldots, 12\}$, $\{13, 14, \ldots, 24\}$,...$\{49, 50, \ldots, 59\}$ must contain at least two of them. Suppose it is the first one, $\{1, 2, \ldots, 12\}$: split these into four triples, A= $\{1, 2, 3\}$, B= $\{4, 5, 6\}$, C= $\{7, 8, 9\}$ and D= $\{10, 11, 12\}$, and consider the six tickets that consist of AB, AC, AD, BC, BD, CD. It is easy to check that every pair of numbers among the list appears in at least one of those six tickets. Parallel arguments hold for the other three lists of a dozen numbers, and, a fortiori, if six tickets suffice for a dozen numbers, six will certainly suffice for the last set $\{49, 50, \ldots, 59\}$ of eleven numbers, so 30 carefully chosen tickets will guarantee a win of some sort. (Can you do better?)

Which is more likely, in the current 6/59 format, that the six winning numbers contain, or do not contain, a consecutive pair such as 26 and 27? (Most

people's first thought is that it is *far* more likely that there is no such pair.)
Let $y_1 < y_2 < y_3 < \cdots < y_6$ be chosen from the list $\{1, 2, \cdots, 54\}$, and write
$x_i = y_i + i - 1$, for $i = 1, 2, \ldots, 6$. By construction, the x-numbers belong to
$\{1, 2, ..., 59\}$, and none of them are consecutive; and given any set $\{x_i\}$ of six numbers from the list $\{1, 2, ..., 59\}$, none of which are consecutive, writing $y_i = x_i + 1 - i$
gives six different numbers from $\{1, 2, ..., 54\}$.

We have set up a one–one correspondence between the collections of six numbers
from 54, and the collections of six of the first 59 numbers, none consecutive, so the
size of each of these sets is the same, and is thus

$$\binom{54}{6} = 25,827,165 = M,$$

say. That means that $N_2 - M = 18,230,309$ of the possible choices of six numbers
from 59 *do* contain a consecutive pair—about 40%. Yes, no consecutive pair *is* more
likely, but not by an overwhelming margin.

This gives a hint of skill in this game of chance. Because many people confuse a
random scatter of numbers with a *uniform* scatter, spread evenly around the ticket,
if they seek to *construct* a 'random' selection of numbers, then far more often their
choice will *not* contain a consecutive pair. So if your selection does contain a consecutive pair, it makes no difference to your winning chance—all N_2 choices are taken
as equally likely—but, if you do fluke the six winning numbers, you are more likely
to share the jackpot with *fewer* other people, and so win more money. See Riedwyl
(1990) or Henze and Riedwyl (1998), for which access to data on the number combinations that punters had chosen in the Swiss Lottery was given. (This information
about the UK Lottery is not generally available—'commercial confidentiality' is
cited.)

Henze and Riedwyl found that punter choice was far from random! Many chose
the winning numbers for the previous lottery or lotteries (possibly on the grounds
that 'nobody else will think to do that'); selections forming horizontal, vertical or
diagonal lines on the ticket to mark the choices were remarkably popular, and, as
noted above, selections with 'clusters' of numbers tended to be avoided. And because
many gamblers used birth dates of family and friends, *lower* numbers were chosen
more often than higher ones.

By and large, a combination will tend to be selected by fewer players than average
if the numbers

  (i) are biased towards higher values;
 (ii) fall into little clusters, rather than being evenly scattered;
(iii) include some on the edge of the ticket;
 (iv) form no obvious pattern on the ticket;
  (v) have not been a previous winning combination.

So if you must buy Lottery tickets, then, to avoid copying someone else's thought
process, try to make a genuine random choice: use some auxiliary device—maybe
take an ordinary pack of 52 playing cards, add seven more from a second pack to

give 59 cards altogether, to be identified with the 59 available numbers. Shuffle them well, deal out six of them, but, unless their numbers meet the criteria above, reject them, return them to the pack, shuffle and deal again.

The Euromillions game is similar, but asks gamblers to choose 5 numbers from 50, *and* 2 'Lucky Stars' from 11. The number of different combinations is thus

$$\binom{50}{5} \cdot \binom{11}{2} = 116,531,800,$$

so the chance of a jackpot share, the reciprocal of this, is less than 10^{-8}, less than half as likely as the UK National Lottery (but with the prospect of much higher prizes). You will win something if you match at least two main numbers, or one main number and both Lucky Stars, 13 different ways in all. See Exercise 7.2.

We might term the Euromillions format as 5/50+2/11; it is apparent from these two examples that, when the Lottery organisers have decided a ballpark figure for the chance of winning the top prize, there will be a variety of ways of tweaking formats like the National Lottery's r/K, or the Euromillions $r/K + a/b$ to achieve their aim. Ian Walker (Chap. 22 in Hausch and Ziemba 2008) points to the attraction of *rollovers*—if the jackpot prize is not won, its value is added to the sums available in the next Lottery—in relieving the tedium of the standard game. (The change in the UK Lottery, from 6/49 to 6/59, by trebling the number of different tickets, has led to many more rollovers.)

If, in any of these lotteries, you wish to buy r tickets, with $r \geq 2$, how do your prospects compare if all the number combinations you select are different, or if you use the same selection every time? Let J be the size of the jackpot that will be generated, and take N as the number of different tickets possible, all equally likely to win.

Suppose you buy r different tickets, and the respective numbers of other punters who have also bought these tickets are $\{x_1, x_2, \ldots, x_r\}$. If the ith ticket happens to hit the jackpot, you will win $J/(x_i + 1)$, so the average amount you will win can be written as

$$\sum_{i=1}^{r} \frac{J}{N(x_i + 1)};$$

similarly, if you buy r copies of the same ticket, your average winnings will be $J.r/(N(x + r))$, where x is the number of other punters who also bought this ticket. You cannot know x, or the values x_i, so it is possible for either of these expressions to exceed the other. But the typical values of x_i or x are low single-digit numbers, so it is very likely that, with $r \geq 2$, $1/(x_i + 1)$ will exceed $1/(x + r)$. Thus we expect $\sum(1/(x_i + 1))$ to exceed $r/(x + r)$, so buying different tickets is a *superior strategy* to spending the same sum on multiple copies. If just one other person had bought the same tickets as we did, having two chances of half the jackpot is better than one chance of two-thirds of it.

Whatever the format, it is usually the case that no more than 50% of ticket sales get returned as prizes, so, whatever strategy used, the large majority of players lose money. Some players will only buy lottery tickets when the 'free money' from rollovers, or artificial top-ups, is sufficiently large.

A different format is used in the Numbers Game, popular in many states in the USA: punters choose three or four numbers from the list $\{0, 1, 2, ..., 9\}$, repetitions being allowed, but now the *order* matters. Thus, with one ticket, and ten different digits available at each of three or four places, the chance of the top prize is $1/1,000$, or $1/10,000$ respectively.

In the most common format of Keno, fresh draws are made every 4–5 min: the Lottery selects 20 numbers from a list of 80, punters may select just 1 or 2, or, if ambitious, up to 10 or even 12 numbers, winning prizes for (say) having at least 5 winning numbers among the 10 they chose. The mean return varies considerably in the different States where this game is played: for example, when selecting 10 numbers with prizes awarded when at least 5 winning numbers have been chosen (or none at all!), it ranges from 49% in Washington to 69% in Massachusetts.

7.3 Roulette

A European roulette wheel has 37 slots, labelled 0–36; an American wheel has an extra slot, double zero. Zeroes are coloured Green, half the other 36 numbers are Red, the rest are Black, with half of each colour being Odd numbers, and also half of them High numbers (i.e. 19–36). Figure 7.1 shows the standard European wheel.

A casino balances and oils the wheels carefully, and we will take it that all 37 or 38 outcomes are equally likely. If you bet one unit on a single number, the standard payout price is 35–1; thus, on average, over 37 bets on a European wheel, you will part with 37 units, and receive back 36 units the one time you win—your average loss per bet is $1/37$ of your stake. This is a very thin margin, just 2.7%, hence the casino's strong incentive to keep its wheels in prime condition. The margin is the same on bets of pairs of numbers (payout price 17–1), triples (at 11–1), quads (8–1), six numbers (5–1) or dozens (2–1): over 37 bets you will expect to lose one unit. On American wheels, the payout prices are the same, but the extra slot increases the casino's advantage: it takes 38 unit bets, on average, to receive 36 units back, so the average loss is $2/38 = 5.3\%$.

On either wheel, the payout prices for bets on blocks of 18 numbers—Red or Black, Odd or Even, Low or High—are at even money: if you bet one unit, two units are returned on a win. Often Zero loses on all these bets, but sometimes, the 'in prison' rule applies: Zero loses, but you only lose half your stake. To cover the different cases, we write p as the true winning probability on an even money bet, and put $q = 1 - p$. Take it that $0 < p < 0.5$.

Ignoring Zeroes, for each n with $1 \leq n \leq 9$, there are 36 overlapping segments of $2n$ numbers. Since Red and Black alternate, all these segments automatically contain the 'correct' numbers of these colours; however, for reasons similar to those noted

Fig. 7.1 The European
roulette wheel

when considering the standard dartboard in Exercise 3.14, it is impossible for the
same to always occur with Odd/Even or High/Low-some segments must have some
bias. But Percy (2015) notes that the discrepancy within any biased section never
exceeds one misplaced number.

Suppose you enter the casino with an initial fortune of F units, and decide to bet
one unit at a time on some even money chance—Red, say—and will play until your
fortune reaches some higher target $T > F$, or you lose all your money. How likely
are you to reach that target, and how many bets will you make?

Write x_n as the chance that, with a current fortune of n, you hit the target before
you lose all your money, for $0 \leq n \leq T$. Obviously $x_0 = 0$ and $x_T = 1$. For
$0 < n < T$, looking at the two possible outcomes of the next bet, when your fortune
will either increase or decrease by one unit, we find

$$x_n = px_{n+1} + qx_{n-1}, \tag{7.3}$$

a so-called *difference equation*. The standard way to solve such equations should
remind you of the path used in Chapter Two to solve second-order *differential* equations with constant coefficients. (Indeed, highlighting this link between *continuous*
and *discrete* mathematics is a strong motivation for including this section.)

Put $x_n = \theta^n$ in (7.3), giving

$$\theta^n = p\theta^{n+1} + q\theta^{n-1},$$

which collapses to

$$p\theta^2 - \theta + q = 0.$$

This quadratic factorises (recall $q = 1 - p$) to give two different solutions, $\theta = 1$ and $\theta = q/p = R$, say. Now any constant multiple of a solution to (7.3) is also a solution, as is the sum of any two solutions. Thus $x_n = A + BR^n$ is a solution, for any constants A and B. The extreme cases $n = 0$ and $n = T$ show that

$$0 = A + B, \quad 1 = A + BR^T,$$

so the chance of reaching the target from n units initially is

$$x_n = \frac{R^n - 1}{R^T - 1}. \tag{7.4}$$

On an American wheel with $p = 18/38$, then $R = 20/18 = 10/9$, so the chance you double your initial fortune of 10 units before you are bankrupted is found to be $0.2585\ldots$, just under 26%. On a European wheel without the in prison rule, $p = 18/37$ so $R = 19/18$ and the chance of the same event rises to almost 37%; with the in prison rule, then we take $p = 73/148$, so that $R = 75/73$ and the chance rises to over 43%.

A similar approach works to find m_n, the mean number of spins you will play altogether, if your fortune is now n. Here $m_0 = m_T = 0$, and, for $0 < n < T$, considering one spin leads to

$$m_n = 1 + pm_{n+1} + qm_{n-1}. \tag{7.5}$$

Again, mirror the method for differential equations: we look for a *particular solution* to the given equation, as well as the general solution to the homogeneous version— which we have just solved. Try $m_n = \alpha n$ in (7.5); it will be a particular solution so long as

$$\alpha n = 1 + p\alpha(n + 1) + q\alpha(n - 1),$$

giving $\alpha = 1/(q - p)$. The general solution to (7.5) will be

$$m_n = n/(q - p) + A + BR^n,$$

so using the known values of m_0 and m_T we find

$$m_n = \frac{n - Tx_n}{q - p},$$

where x_n is as given above.

On average, hoping to turn 10 units into 20 in this manner, you would play about 92 spins on an American wheel, and 98 on a European wheel, or 99 if the in prison rule applied. Whichever rule applies, the mean number of bets you make is quite similar, but the chance of doubling your money, rather than losing it all, does differ appreciably.

If your sole intention is to double your money, then betting one unit at a time would be poor play: a single bet, of your entire fortune, would achieve your goal with probability p (under the different rules $p = 18/38 = 47.4\%$, $18/37 = 48.6\%$ or $73/148 = 49.3\%$), much larger than the drawn-out play. This illustrates a general principle: in unfavourable games, you maximise your chance of reaching a goal through bold play, expecting to make few bets. Cautious play, hoping to edge your way to a larger sum, will be less successful—but will prolong your time in the casino.

But suppose the casino have been careless, and the wheel is so biased towards certain numbers that there are favourable bets: for example, maybe the true probability of Red is $p > 0.5$—how best to take advantage of this?

The answer is found in Chap. 4, where the Kelly Strategy is described: you should bet the fraction $2p - 1$ of your current fortune, and you can then expect your fortune to grow at the average rate of $2p^p(1 - p)^{1-p}$ per bet. Even if you are lucky enough to find this opportunity, your fortune will not build up quickly—if $p = 0.5 + \epsilon$, with $\epsilon > 0$ small, this growth rate is around $200\epsilon^2\%$. Using the Rule of 72, this indicates that it will take, on average, around $9/(25\epsilon^2)$ bets to double your fortune— with an optimistic value of $p = 52\%$, you should expect to need about 900 bets of 4% of your current fortune to double its initial value!

A more general scenario is described by Thomas Bass (1991) (also published under the title 'The Eudaemonic Pie'). A group of physics students used sensors in their shoes to transmit information on the speed of the ball as it travelled to predict into which region of the wheel it would settle; if their predictions were accurate enough—they did not need certainty, it was enough to know that (say) landing in a particular set of six adjacent numbers was at least $1/6 = 16.7\%$, rather than $6/38 = 15.8\%$, to give them an advantage. To make the best of any advantage, they had also to judge the optimal size of bet. The maths is that of the adaptation of the basic Kelly Strategy, as described in Exercise 4.17. We will illustrate this by looking at the optimal strategy if the chance of some particular single number is $p > 1/36$.

The payout odds on single numbers are $35 : 1$, so we know that bets on that number are favourable: and it transpires that the optimum growth rate of your fortune is achieved if you bet the fraction $x = p - (1 - p)/35$ of your fortune. Your fortune's growth rate then turns out to be around $17.5x^2(1 - 34x)$ per spin. For example, suppose p is as large as $1/30$: then you should bet just over $1/200$th of your fortune each spin, expecting an average growth rate of just under 0.05%, and taking over 1500 spins to double up. This is not a swift or sure route to riches, despite your advantage.

A different idea is to aim to leave the casino with a profit of one unit: to do so, you will wait until three successive spins show Black, and then begin a sequence of bets on Red, of sizes 1, 2, 4, 8, ..., doubling the bet after each loss, and stopping at the first success. If Red arises for the first time on your nth bet, your stake will be 2^{n-1}, and you will have lost a total of

$$1 + 2 + 4 + \cdots + 2^{n-2} = 2^{n-1} - 1$$

to date; so your net position after this winning bet is easily seen to be a profit of $2^{n-1} - [2^{n-1} - 1] = 1$ unit.

This holds whatever the value of n, and Red will certainly appear sometime! But we have not found the Holy Grail, a guaranteed method of winning, for two linked reasons. First, although it is quite likely that Red will appear fairly soon, and we do bank our profit, we cannot ignore the possibility that our funds are insufficient to meet the next bet the system requires; casinos do not allow credit. Second, there will be both a minimum and a maximum permitted bet—perhaps the maximum is 100 times our initial bet. Then, if the first seven bets lose, the last losing bet was of 64 units and the system demands a bet of 128 units, which is not allowed.

If the wheel is fair, and the chance of Red is $18/37$ each time, the chance of losing on seven consecutive bets is $(19/37)^7$, just under 1%. So 99% of the time, we do make our one unit profit, but the loss of 127 units when we fail more than outweighs this. Exercise 7.7 asks you to show that the mean number of bets you make would be is around 2.05, with an average total stake of about 4.3 units.

(You will realise that the only effect of waiting until three consecutive Blacks appear before making your first bet is to prolong your time in the casino: waiting has no effect at all on your chance of winning, or the mean amount of your losses.)

7.4 The Horse Racing Family

Betting on horses has been a substantial activity in the UK and elsewhere for hundreds of years. For each horse, bookmakers will offer a payout price of the form α to 1; this price will reflect both the objective chance that the horse wins the race, and the amounts of money that punters bet on it. Write S as the sum, over all the horses, of the quantity $1/(1+\alpha)$; it is crucial that $S \geq 1$, (and we often find that S substantially exceeds unity), for otherwise an observant punter could guarantee to win money, whatever the result of the race.

For, suppose $S < 1$, and the punter has a fortune of size F. By betting the entire amount, placing $F/(S(1 + \alpha))$ on a horse with payout price $\alpha : 1$, the total return will be F/S whichever horse happens to win—and $F/S > F$ since $S < 1$.

Example 7.1 In a race with four horses, the bookmakers have erred: the payout prices are posted as 3–1, 4–1, 5–1 and 6–1. Suzie has £319 in her purse: she stakes £105, £84, £70 and £60 on those horses, respectively; whichever wins, she will be paid £420, a profit of 32% on her total stake.

The amount by which S exceeds unity is known as the *overround*, or *bookies' margin* or *vigorish*. In the soccer example in the Introduction, with payout prices for the three outcomes as 5–6, 4–1 and 12–5, you should check that the overround is almost 4%. In horse racing, the overround varies with the size of the field, with an average figure of around 20% found by Smith et al (2006). The Grand National, with up to 40 runners, has a *much* larger overround.

As well as betting on a horse to win, it is also possible to bet on a horse being *placed*: this bet is not offered if there are fewer than 5 runners, otherwise the general rule is that placed means first or second in a race with 5–7 runners, in the first three with 8–15 runners, and in the first four with 16+ runners. The place odds are derived from the win odds: usually, the formula is one-quarter the odds (i.e. if the win odds are α to 1, the place odds are $\alpha/4$ to 1), but this often reduces to one-fifth the odds for races with 8–15 runners.

Statistician Robin Plackett (1975) described how a bookmaker had approached him for an explanation of the fact that it appeared to be possible that, although the bookie had set odds with a positive overround for the win market, punters might be able to guarantee a profit by betting on the place market, derived from the win odds as above. See Exercise 7.8.

A different betting format is the Tote, or pari-mutuel system. Suppose that, in a race with n runners, punters collectively stake amounts $\{x_i\}$ on the ith horse to win, with $T = \Sigma x_i$ the total amount staked on that race. A proportion p of this amount is deducted to cover the running costs (and profits) of the Tote, leaving $T(1 - p)$ as the *net pool*, which is divided among the winners in proportion to the amounts they staked—exactly how the higher prizes in many Lotteries are determined. Formally, if horse i is the winner, then the *dividend* is $D = (1 - p)T/x_i$, (rounded to the nearest $10p$ below); a punter who had staked amount A on this horse will receive $A.D$.

The same principles apply to other types of pari-mutuel bet, such as the place market, the Exacta (forecast the first two horses in the correct order), the Trifecta (get the first three in the correct order) and more exotic bets. In the UK, the amount of the deduction p to form the net pool is currently 16.5% for the winner, 18% for places and up to 30% for other bets. In some countries, pari-mutuel is the only legal format.

If the net pool for the place market is P, and the first r runners qualify, then the amount P/r is allocated to each 'place', and the backers of that horse similarly receive an amount proportional to their own stake. Exercise 7.9 points to a problem with this setup.

In the 1973 Belmont Stakes, the winner was Secretariat, such an overwhelming favourite that the return on a 2-dollar bet to win was $2.20 only. But shrewd punters who put their 2-dollar stake on Secretariat being placed (in the first two horses) obtained a return of $2.40! (We have noted that similar anomalies have arisen in the UK National Lottery.)

From a punter's perspective, the main difference between bookmakers and the Tote is that, with the Tote, they cannot know in advance the size of their potential winnings—it depends on how much other people have staked on the same horse. But continuous guidance is available, on screens that display what the Tote payouts would be, updated after every bet placed: you can watch your own bet shift those amounts (if it is large enough).

Example 7.2 Suppose a race has 12 horses, all at bookies' odds of 9–1 to win, giving an overround of 20%. Suppose also that the total win stakes at the Tote across all horses are £120,000, and that the winner is Gullible, with a Tote total stake of

£16,700. The net pool is 83.5% of total stakes, i.e. £100,200, so the Tote dividend is 100,200/16,700 = 6.0. A punter who had bet £10 on Gullible to win would receive back £60 had the stake been with the Tote, but £100 for a bet with the bookies.

But if the winner had been Optimistic, with a total Tote stake of £8350, the Tote return would have been £120, with £100 again from the bookies. In this example, there is little overall difference between the two systems—if £120,000 were staked with the bookies, the actual amount they paid out would vary with the winner, but the *average* total payout would be £100,000, comparable to the *certain* total payout of just £200 more with the Tote.

Note that *Football Pools* operate along the same lines as the Tote. The Pools promoters deduct a percentage of stakes to cover their costs and to provide profits, and return the rest to winning punters, in proportion to the amount they have staked.

7.5 Card Games

There is no such animal as a 'skilled baccarat player'. Baccarat is a game of pure chance, the actions of dealer and player being constrained by rigid rules. But most card games do provide opportunities for skill to be shown, and proficient *counting* is usually the key.

Poker has many variant forms, but the basic rules are simple: each player will have a hand consisting of five cards, and the one with the best hand wins. The best possible hand is a Straight Flush—five cards of the same suit, and with consecutive ranks: then come Four of Kind (four cards of the same rank, plus one other card), Full House (three cards of one rank, two of another), Flush (five of the same suit, not consecutive), Straight (five consecutive cards, not all the same suit), Three of a Kind (three cards of one rank, two other cards of different ranks), Two Pairs (two cards of one rank, two of another, one other card), One Pair (two cards of the same rank, three other cards of different ranks) and finally High Card (none of the above). Within each category, it should be clear which of two hands is the better.

The number of possible hands is $\binom{52}{5} = 2,598,960$. Since a Straight can begin (working upwards) with any one of the ten possible ranks Ace to Ten (as Ace can count both high or low at your choice), and there are four suits, there are $10 \times 4 = 40$ Straight Flushes. With 13 different ranks, and then 48 ways to select the last card, we find $13 \times 48 = 624$ hands forming Four of a Kind. For a Full House, there are 13 ways to choose the rank for the triple, and, within that, 4 ways to select those 3 cards; combine these with 12 ways to choose the rank of the pair, and then 6 ways to choose which two suits form that Pair, leading to $13 \times 4 \times 12 \times 6 = 3744$ hands in all.

There are $\binom{13}{5} = 1287$ ways to choose 5 cards from any suit: but 10 of these Flushes are Straight Flushes, leaving 1277 ordinary Flushes in each suit, so 5108 altogether. With 10 ways to begin a Straight, and 4 suit choices at each stage, there are $10 \times 4^5 = 10,240$ combinations, but 40 of these are Straight Flushes, giving 10,200 ordinary Straights.

Table 7.2 Types of poker hand, and their respective numbers

Type of hand	Number of ways
Straight flush	40
Four of a kind	624
Full house	3744
Flush	5108
Straight	10,200
Three of a kind	54,912
Two pairs	123,552
One pair	1,098,240
High card	1,302,540

Table 7.3 Types of hand, and typical payout odds at Video Draw Poker

Type of hand	Payout odds
Royal flush	799:1
Other straight flush	49:1
Four of a kind	24:1
Full house	8:1
Flush	5:1
Straight	3:1
Three of a kind	2:1
Two pairs	Evens
One pair, Jacks+	Stake returned
Worse than two Jacks	Loss

Exercise 7.10 asks you to make the appropriate calculations for the other types of hand, and thus verify the values in Table 7.2.

Perhaps the simplest form of this game is Draw Poker: each player is dealt a poker hand, and may choose to discard one or more cards, replacing them with an equal number from the remaining pack. The player with the best hand wins.

The four Straight Flushes that consist of Ten to Ace are termed Royal Flushes, and are especially valuable in Video Draw Poker machines found in casinos: typically, unless the final hand is Two Jacks, or better, the stake is lost. Plainly, just 4/13 of One Pair hands are indeed at least as good as Two Jacks. Hands are paid out at odds similar to those displayed in Table 7.3.

Michael Shackleford, who writes under the name 'The Wizard of Odds' is a reliable authority on this and many other games found in casinos. Different machines have subtly different rules and payout odds, but the Wizard has calculated the best strategy for all the machines with which he is familiar, and generously displays his findings for free on the internet.

Example 7.3 Based on the payout odds of Table 7.3, what are the plausible strategies, and which would give the best mean return, if you are dealt $\{K, Q, 7, 5\}$ of Spades and the Five of Hearts?

Solution. Since we have two choices with each of five cards, keep it or reject it, there are $2^5 = 32$ possible strategies, but just three of them are plausible. The simplest, (a), is to keep the four Spades and discard the Heart, hoping for a Flush or High Pair. Of the 47 cards to be drawn from, nine are Spades and six are King or Queen, so the chance of a Flush is 9/47, of a High Pair it is 6/47. Using Table 7.3, we see that that the mean return is $(9/47) \times 6 + (6/47) \times 1 = 60/47$.

Alternative (b) is to keep both Fives, discarding the other three Spades. A pair of Fives is worthless, but it opens up several possibilities. Exercise 7.11 asks you to verify the calculations for this strategy, and also alternative (c), which is to keep just the King and Queen of Spades, giving chances of a Royal or Straight Flush. In both cases, there are $\binom{47}{3} = 16,215 = N$, say, ways to select the other three cards.

With (b), drawing both the remaining Fives and one other card gives 45 chances of Fours; there are 165 ways to get a Full House, 1854 for Threes and 2592 choices that give Two Pairs. All other choices, 11,559 in number, are losing hands, so the mean return is

$$(45 \times 25 + 165 \times 9 + 1854 \times 3 + 2592 \times 2)/N = 13,356/N \approx 0.824.$$

For (c), there is just one way to obtain a Royal Flush, and one way to obtain an ordinary Straight Flush, while drawing the remaining three Kings or Queens gives two ways for Fours. There are 18 ways to get a Full House, 82 for a Flush and 126 for a Straight. (A little harder) we find 283 ways to get Threes, 717 to get Two Pairs and 5016 to get a High Pair, leaving 9969 losing selections. The mean return is

$$(800 + 50 + 2 \times 25 + 18 \times 9 + 82 \times 6 + 126 \times 4 + 283 \times 3 + 717 \times 2 + 5016)/N,$$

approximately 0.577. The choice with highest mean return is (a), discard the Five of Hearts.

Plainly, it is impractical to make these calculations while seated at a machine in the casino; print off a copy of the Wizard's relevant cheat sheet.

Another version of poker found in casinos is Caribbean (or Casino) Stud Poker. There are minor rule variations but, in all formats, the Dealer plays on behalf of the casino against up to six Players. Initially, each Player selects his basic stake—his Ante, and all are dealt five cards from a single freshly shuffled deck, the Dealer's final card—her Upcard—is exposed.

On seeing his cards, a Player will Fold—i.e. drop out and concede his Ante—or Raise: a Raise is an additional stake of twice the Ante. When all decisions are made, the dealer exposes the rest of her cards. If she does not have a hand at least as good as High Card with Ace and King (a non-Qualifying Hand) only the Antes are in play: those who did not Fold win just the amount of their Ante. If she does have a Qualifying hand, she matches it against each player who made a Raise. A winning player wins the amount of his Ante, plus a Bonus on his Raise that depends on how good his hand is; for example, a winning Straight Flush is paid out at odds of 50:1, a winning Two Pairs hand is paid out at 2:1. A losing hand just loses Ante and Raise.

When should a Player Raise? If he Drops, he loses his Ante, one unit, so he should Raise when, on average, the outcome from doing so exceeds a loss of one unit. After a considerable amount of well-organised counting, in which the Upcard plays a very minor role, you will find that you should always Drop if you have a non-Qualifying hand, and always Raise if your hand is AKQJx High Card or better. For the small proportion of hands that do not fall into one or other of these categories, see either Griffin and Gwynn (2000) or Haigh (2002). To a good approximation, if you hold a hand at least as good as a Pair of Sevens, you will make a profit, on average. And based on the average stake if using optimal play, the casino's edge is about 2.36%, slightly smaller than in roulette.

Currently, the most popular poker format outside casinos is Texas Hold'Em. Each player has two cards, but five other cards are dealt face up on the table, and players seek to make the best possible hand via some combination of their own cards, and the five communal cards.

I will not detail all the rules. But it is important to know that there are up to four rounds of betting: after your two cards are dealt; after the first three communal cards (the 'flop'); after the next card (the 'turn'); and after the final card (the 'river'). At each stage, you must decide whether to retire gracefully, abandoning your cards and the amounts you have staked, or match other people's bets, or even raise the stakes. In contrast to Video machines in casinos, you are now facing human opponents who will try to outwit you, either by intimidating you into retiring by making big bets (without holding cards that objectively justify their actions), or by sandbagging—luring you on with small bets, perhaps hesitant plays, while actually holding an unbeatable hand. Mathematical calculations help, but psychological factors are important.

In evaluating your hand, the number of players in the game is important. The Wizard's website is a mine of useful information; he gives the answers, but not the calculations that lead to them. It is instructive to peruse these answers, and to think why they hold. For example: always, your best possible two-card holding is a Pair of Aces, which will win nearly 85% of the time with just two players, and over 31% of the time with ten players; with six to ten players, the *worst* hand is Seven and Two of different suits, but with four or fewer players, the worst hand is Three and Two of different suits.

Phil Woodward (2006) attributed the following quirky example to 1972 World Champion Amarillo Slim. Consider the three two-card hands:

(a) Two of Clubs, Two of Spades;
(b) Ace of Spades, King of Diamonds;
(c) Jack and Ten of Hearts.

Slim was happy to allow his opponent to make a choice from those alternatives, then he would pick one of the remaining hands, and play one game of Texas Hold'Em for $1,000. He expected to win, in the long run, since calculations show that (a) beats (b) some 52% of the time, (b) beats (c) 59% of the time, while the chance (c) beats (a) is about 53%!

(The issue of Significance that contains Woodward's article includes many other pointers to mathematical and statistical aspects of gambling.)

7.6   Premium Bonds

Premium Bonds have been on sale in the United Kingdom since 1956. They are issued by National Savings and Investments, backed by HM Treasury. An individual may hold up to 50,000 Bonds, each costing £1, and (as of September 2019), every month, each Bond has one chance in 24,500 of winning a tax-free prize. These parameters have changed over time, but this broad framework has prevailed.

It has been argued that, since any Bond can be redeemed at its full cost price, without notice, this is not a gamble, as the Bond-holder cannot lose; but the prize fund is generated by a notional interest rate, currently 1.40%, so the Bond-holder foregoes the chance to earn interest directly. Assuming that annual interest of 2.5% would be paid to someone depositing £10,000 with a bank for three years, buying Premium Bonds instead exchanges the certainty of £750 interest for 10,000 chances in each of the next 36 months to win prizes. The average amount won would be £420, so we suggest that that this Bond-holder has gambled £750, with a mean return of 56%.

The total number of Bonds currently held is about 80 billion, and the prize amount generated each month is over £90 million. This funds two prizes of one million pounds, and over three million prizes of lower sums: about 90% of the prize money is allocated to prizes of £100 or less. Indeed, by number, over 98% of the prizes are for the minimum amount, £25. Like the National Lottery, the prize structure gives a remote chance of an enormous prize, and a respectable chance of some sort of prize.

How remote? Even if you hold 50,000 Bonds, the chance of winning a million pounds in any month is about 1 in 800,000, and the average time you would wait for the top prize is over 60,000 years! But, with that holding, the chance you will win *something* in any given month is about 87%.

The winning numbers are generated by ERNIE (Electronic Random Number Indicator Equipment). The original machine was built by members of the famous Bletchley Park code-breaking team. It has been replaced several times; the current version, ERNIE 5, is over 21,000 times as fast as the original machine, and completes the draw in around twelve minutes. A description of the tests made on the output to enable the Government Actuary's Department to certify the draw is satisfactory, can be found in Field et al. (1979).

The original ERNIEs used 'thermal noise' machines to produce random numbers; this new machine uses quantum technology, based on light. Emphatically, ERNIE has never worked in the same way as pocket calculators, or modern digital computers, which are programmed to produce 'pseudo-random' numbers—these numbers do pass stringent tests of 'randomness', but, as described in the next chapter, are actually produced by an arithmetical formula which will give the same sequence again and again, given the same initial value.

7.7 Exercises

7.1 In the UK National Lottery, as well as the six winning numbers, a seventh one (the Bonus Ball) is also chosen: punters whose selection consists of any five winning numbers, along with the Bonus Ball number, qualify for the second prize, after the jackpot, those with five winning numbers but not this Bonus number qualify for the third prize. In the old 6/49 format, prizes were also given for matching three or four numbers, in the new 6/59 format there are prizes for matching two, three or four numbers. For each format, give the chances, to three significant figures, of winning at each prize level.

7.2 This Chapter notes that there are thirteen ways of winning a prize in the Euromillions game. List these different ways, count how many different gambler choices will give each of them, and hence give their odds (appropriately rounded to 1 in M, for suitable integers M).

7.3 Suppose you are in Las Vegas with $100, which you hope to turn into $400 through roulette on an American wheel with 38 numbers. Find your chances of doing so, if you confine yourself to bets on Black, with constant stakes of (i) One dollar (ii) $10 (iii) $50 (iv) $100.

7.4 Suppose you find a casino with a biased wheel, so that the chance of Red is some value $p > 1/2$. Your begin with fortune F and bet on Red, one unit at a time, until either you have a run of bad luck, and lose everything, or your fortune builds indefinitely (the casino has limitless resources). Use the fact that, in the expression $x_n = (R^n - 1)/(R^T - 1)$, we now have $R < 1$, to find the chance you are ruined before the casino.

7.5 Suppose $p = q = 1/2$ in Eq. (7.3). Then, in parallel with *differential equations*, when the auxiliary equation has equal roots, the general solution is $x_n = A + Bn$. Deduce that $x_n = n/T$, and use L'Hôpital's Rule to verify that, as $R \to 1$ in Equation (7.4), so $x_n \to n/T$.

Similarly, find the new expression for m_n, the mean number of spins until ruin or triumph.

7.6 In roulette, Isabella will bet one unit on her favourite number until her money runs out: the payout odds are 35 to 1, her winning chance on any spin is 1/37. Find the chance that she is *ahead* after n spins when (a) $1 \le n \le 35$ (b) $37 \le n \le 71$ (c) $73 \le n \le 107$. At which point is she more likely to be ahead, after 175 spins, 179 spins or 181 spins?

7.7 Suppose that, as described in the text, you bet one unit initially on Red, double up your bets after a loss, and quit either when you have won, or you make N consecutive losses, whichever comes first. Let $p < 0.5$ be the chance of Red on any spin, and write $q = 1 - p$. Show that the mean number of bets you make will be $(1 - q^N)/p$, and the mean total stake will be $(2^{N-1}q^N + q - 1)/(2q - 1)$; verify the figures given in the text, by evaluating these quantities when $N = 7$ and $p = 18/37$.

7.8 Consider a horse race with 16 runners, with the payout odds for each horse to win being α to 1, and the payout odds for a place being $\alpha/4$ to 1, 'place' here meaning finishing in the first four, as described in the text. What condition on α implies a positive overround for the bookies on the win market? For what values of α can the bookies have a positive overround on the win market, but a shrewd punter could guarantee to win on the place market?

If the bookies sought to protect themselves by insisting that any bet on a horse to place must be accompanied by a bet of the same size on that horse to win, would it still be possible for a shrewd punter to guarantee a profit?

7.9 Suppose the net pool for the place market in a ten-horse race is £150,000, and the totals staked on the first three horses are £100,000, £10,000 and £20,000. Derek has bet £10 on each of these, so has three winning bets. According to the formula given in the text, show that he should receive £5, £50 and £25, respectively.

(So Derek staked £10 on the horse that finished first, the bet was successful, but the formula gives him £5 only—a *loss* of £5. In practice, to avoid this absurd outcome, special rules are applied to ensure that such 'winners' do not receive less than their stake. One such method is described by Skiena (2001): *first*, allocate to each place pool the total amount staked on that horse, *then* divide the rest of the place pool into equal amounts, and add this sum to each pool.)

7.10 Make the calculations that verify the entries in Table 7.2.

7.11 Make the calculations that verify the mean returns for strategies (b) and (c) in Example 7.3.

7.12 In Casino Stud Poker, show that there are 1,135,260 non-Qualifying hands, about 43.7% of the total. Consider the five hands R= $\{J, 9, 8, 5, 2\}$, S= $\{A, K, J, 10, 9\}$, T= $\{A, A, J, 4, 3\}$, U= $\{6, 6, 6, J, 4\}$ and V= $\{J, J, 8, 8, 8\}$, all non-Flush. Decide on which hands you would Fold or Raise; on those decisions, what would the outcome be if the Dealer held (i) Hand R; (ii) Hand S; (iii) Hand U?

7.13 Watch a clip of the 2006 film 'Casino Royale', showing the final hand of 'Texas Hold 'Em', where the five communal cards are Spades $\{A, 8, 6, 4\}$ and the Heart Ace. The four remaining players held, concealed, (a) Spades $\{K, Q\}$, (b) $\{8, 8\}$, (c) $\{A, 6\}$ and (d) Spades $\{7, 5\}$. On what each player could see, how many of the $\binom{45}{2} = 990$ possible holdings for an opponent would beat their hand? Were they all justified in going 'all-in'?

7.14 Use the information given in the text to verify the claims made for holders of 10,000, or 50,000, Premium Bonds. As the prizes are tax-free, what is the interest rate that a person whose marginal tax rate is 40% would need to obtain on a bank deposit, to match the average return on Premium Bonds?

7.15 Explain (briefly) why it is vital for Premium Bonds that, as described in the final sentence of the text, ERNIE does *not* produce its winning numbers in the same way as standard random number generators in computers or pocket calculators.

References

Bass T A (1991) The Newtonian Casino. Penguin

Field J L, Johnston E A and Poole J C (1979) The Mathematics of Premium Savings Bonds. *Bulletin of the IMA* 15(5/6) pages 132–8

Griffin P A and Gwynn J M Jr (2000) An analysis of Caribbean Stud Poker. In "Finding the Edge: Mathematical and Quantitative Analysis of Gambling" editor Judy Cornelius.

Haigh J (2002) Optimal strategy in casino stud poker. *The Statistician* 51 pages 203–13

Hausch D B and Ziemba W T (2008) Handbook of Sports and Lottery Markets. Elsevier

Henze N and Riedwyl H (1998) How to Win More. A K Peters

Percy D (2015) A New Spin on Roulette: Combination Bets and Unbiased Wheels. preprint

Plackett R L (1975) The analysis of permutations. *Applied Statistics* 24 pages 193–202

Riedwyl H (1990) Zahlenlotto. Wie man mehr gewinnt. Paul Haupt

Skiena S S (2001) Calculated Bets. Cambridge University Press

Smith M A, Paton D and Vaughan Williams L (2006) Market Efficiency in Person-to-Person Betting. *Economica* 73(292) pages 673–689

Woodward, P (2006) Call my bluff. *Significance* 3(1) pages 30–32

Chapter 8
Computer Applications

8.1 Introduction

When commercial computers began to be manufactured soon after the end of World War Two, they had very limited storage space. Today, storage capacity is not an issue, and computers can work speedily and exactly with integers of any size. In the applications we look at in this chapter enormous numbers, often prime numbers, are involved. That there are infinitely many prime numbers was known by Euclid, but the first good estimates of $\pi(M)$, the number of primes that do not exceed M, were given by Legendre around the end of the eighteenth century. The Prime Number Theorem, which states that the ratio of $\pi(M)$ to $M/\log(M)$ converges to unity as M increases without bound was proved a century later by Hadamard and de la Vallée Poussin. Table 8.1 (adapted from Churchhouse 2002) illustrates this approximation.

There is no shortage of very large prime numbers. For example, $\pi(10^{90})$ is estimated to be about 4.83×10^{87}, and $\pi(10^{91})$ to be about 47.72×10^{87}, so between 10^{90} and 10^{91}, there are over 4×10^{88} primes to choose from. About $1/\log(M)$ of the numbers near M are prime, so almost 1 in 200 of numbers near 10^{90} are prime. We make the assumption that we can find very large prime numbers for the purposes we need.

But to illustrate the methods that use these enormous numbers, our examples will, of course, use reasonably small numbers (and you should follow through the accompanying arithmetic).

. The idea of arithmetic modulo m, as seen in Chap. 4, is very useful. Recall that 'm modulo n', often abbreviated to m (mod n), means the remainder r in the range $0 \leq r \leq n - 1$ when the integer m is divided by the integer n. Thus 58 (mod 7) = 2, and -13 (mod 17) = 4. Especially where we look at Codes and Ciphers, this notion is central, and the following result often simplifies the arithmetic.

Theorem *Suppose n and k are positive integers, and that x and y are also integers. If $x = y$ (mod n), then $x^k = y^k$ (mod n).*

© Springer Nature Switzerland AG 2019
J. Haigh, *Mathematics in Everyday Life*,
https://doi.org/10.1007/978-3-030-33087-3_8

Table 8.1 Frequencies of prime numbers

Value of M	No. of primes $\leq M$	$M/\log(M)$
10^3	168	144.8
10^4	1229	1085.7
10^5	9592	8685.9
10^6	78,498	72,382.4

Proof When $x = y$ (mod n), their difference $x - y$ is an integral multiple of n, so $x = y + Rn$ for some integer R. The binomial expansion gives

$$x^k = (y + Rn)^k = y^k + \text{ other terms, all multiples of } n.$$

Hence $x^k - y^k$ is a multiple of n: the result follows.

Example 8.1 We wish to find 13^{23} (mod 51). Modulo 51 in each computation,

$$13^2 = 169 = 16, \text{ so } 13^4 = 16^2 = 256 = 1.$$

Hence $13^{4m} = 1$ for any integer n; in particular, $13^{20} = 1$. Thus $13^{22} = 1 \times 16 = 16$, so $13^{23} = 16 \times 13 = 208 = 4$.

Given any real number x, except $x = 0$, there is a unique number y such that $x \times y = 1$; we say that y is the *inverse* of x, and write $y = 1/x$. There is a parallel in modular arithmetic: if n is prime, then every number x with $1 \leq x \leq n - 1$ has a unique inverse y, in the sense that $x \times y = 1$ (mod n). But if n is not prime, only those numbers that have no factors in common with n have such an inverse.

Example 8.2 It is easy to check that the respective inverses of $(1, 2, \ldots, 10)$ (mod 11) are $(1, 6, 4, 3, 9, 2, 8, 7, 5, 10)$. But modulo 10, only $(1, 3, 7, 9)$ have inverses (they are $(1, 7, 3, 9)$). To see why the rest do not have inverses, note that, for example, $2 \times 5 = 10 = 0$ (mod 10). So if J were an inverse of 5, we would have the contradiction

$$0 = 0 \times J = (2 \times 5) \times J = 2 \times (5 \times J) = 2 \times 1 = 2.$$

Not long ago, humans could smugly write 'Computer—fast, accurate, stupid. Man—slow, slovenly, brilliant'. But now, while man is still comparatively slow and slovenly, computers have not only got far faster, still with complete accuracy, but also by training on millions of games, a computer has 'learned' to play the game GO sufficiently well to beat top-ranked humans. Shall we soon have to conclude that computers are indeed fast, accurate and brilliant?

8.2 Pseudorandom Numbers

By the phrase 'U(0,1) random numbers', we mean a sequence $(u_1, u_2, u_3, \ldots)$ of real numbers in the range (0, 1) such that, given any values a and b with $0 \leq a < b \leq 1$, the chance that a particular member falls in the interval (a, b) is equal to its length, $b - a$, independently each time. ('Independence' has a precise meaning: informally, it means that information about some members of the sequence does not help in forecasting other members.)

As well as 'serious' work, computers provide much entertainment. Players can try their luck at roulette, or play games such as backgammon against another human, or against a computer program. Write INT(X) to mean the largest integer that does not exceed X; then, given a sequence $(u_1, u_2, u_3, \ldots)$ of $U(0, 1)$ values, taking $y_i = $ INT($37 \times u_i$) can mimic roulette, as all 37 outcomes of a spin of the wheel will then be equally likely, no matter what the previous history. Similarly, with dice games, we can use the instruction $y_i = 1 + $ INT($6 \times u_i$) to generate the numbers one to six at random.

In Chap. 3, we noted that when the Kicker takes a penalty at soccer, he may wish to be able to select Left or Right with some respective probabilities p and $1 - p$, independently each time. From a suitable supply of $U(0, 1)$ values, kick Left if and only if today's number does not exceed p.

In a clinical trial, either to test a new treatment against the existing standard, or to test a new drug against a placebo, it is important that patients are allocated *randomly* to one treatment or the other. It cannot be left to individual doctors to toss a coin to decide which arm of the trial applies to the next patient. There must be some overall controller, who is genuinely able to assure the authorities that a proper random allocation has been made. Similarly, Opinion Polls try to select their interviewees at random. And the Times newspaper allocates a different number to each email entry of its crossword puzzles, in order to choose the prizewinners at random.

Digital computers do not generate genuinely $U(0, 1)$ numbers, rather they produce a sequence $(v_1, v_2, v_3, \ldots)$ that *looks as though* it is random, in the sense that it will pass a battery of statistical tests on frequencies, and lack of association between different members. There is some initial *seed* from which successive members of the sequence are found using an arithmetical formula. Given the same seed, the same sequence will ensue, hence the term *pseudorandom sequence*.

Example 8.3 Morgan (1984) examined the scheme, where v_{n+1} is taken as the fractional part of $(\pi + v_n)^5$. For any v_n in the range zero to unity, $3.141 < (\pi + v_n) < 4.142$, so $305.7 < (\pi + v_n)^5 < 1219.2$. With such a wide range of possibilities, it looks reasonable to hope that the fractional part of $(\pi + v_n)^5$ does not favour any zone of the interval (0, 1) over another of the same size, and that the stream of outcomes will appear unrelated to each other. Morgan took an initial value v_0 and listed just the first three significant figures of the values (v_n) as possible *random digits*, i.e. integers in the range zero to nine. Forty iterations yielded 120 digits in roughly equal numbers, and no apparent way to use the past to forecast the future.

The most common methods rely on using a *linear congruential generator*, i.e. given three fixed integers $a > 0$, b, $m > 0$ and some integer seed x_0, obtain the sequence (x_n) recursively via

$$x_{n+1} = ax_n + b \,(\text{mod } m). \tag{8.1}$$

Taking $v_n = x_n/m$ gives a value in the range zero to unity and, if a, b and m are suitably chosen, the sequence (v_n) will pass a range of statistical tests of randomness.

Moreover, unlike the ad hoc method of Example 8.3 and the long-discredited mid-square method of Exercise 8.1, we can hope that mathematics can give some theoretical justification for these generators—to be able to say why particular choices of (a, b, m) in (8.1) should definitely not be used, while other choices can be expected to have the properties we desire.

Since x_n is restricted to the integers $(0, 1, 2, \ldots, m - 1)$, the pattern repeats after at most m iterations. Some applications demand an enormous quantity of values (v_n), so m must be large. Several sets of conditions on (a, b, m) that will ensure a cycle of maximum length exist: perhaps the simplest is that m be prime, and that $0 < b < m$. Division can be relatively slow, but with a computer that works with binary numbers, taking $m = 2^k$ means that the step '(mod m)' is accomplished immediately. With that choice, we would get a full cycle whenever b is odd, and $(a - 1)$ is a multiple of four. With $b = 0$ and $m = 2^k$, the cycle length is 2^{k-2} when x_0 is odd and $a = 8c \pm 3$ for some integer c.

It is not enough to have a long cycle: we desire no discernable association between members of the sequence. One measure of association is *correlation*, a value in the range $(-1, 1)$. With genuine random numbers, the correlation between any pair is zero, so we hope that with pseudorandom numbers, the correlation is near to zero. It turns out that the average correlation between consecutive pairs when using (8.1) is approximately

$$\rho \approx \frac{1}{a} - \frac{6b}{am}(1 - \frac{b}{m}) \pm \frac{a}{m}. \tag{8.2}$$

This indicates that a should also be large, but relatively small compared to m—perhaps a might be approximately $\sqrt{m}$?

There can be hidden pitfalls. Taking $a = 65,539$, $b = 0$ and $m = 2^{31}$ can give a cycle of length 2^{29}, and the average correlation between successive values is less than 10^{-4}. But, since $a = 2^{16} + 3$, we see that, modulo 2^{31},

$$(x_{n+2} - 6x_{n+1} + 9x_n) = x_n((2^{16} + 3)^2 - 6(2^{16} + 3) + 9) = 2^{32}x_n = 0.$$

Consecutive triples are closely related! This generator, used under the name RANDU, was a disaster. See also Exercise 8.2, which illustrates how the cycle length can vary dramatically, according to the choice of seed.

Knuth (1981) noted that (8.2) is an average over a complete cycle; the fact that this average is near zero implies nothing about its behaviour over a short range. With $m = 2^{35}$, $a = 2^{18} + 1$ and $b = 1$, the average correlation is around 2^{-32}, but batches of 1000 consecutive values gave correlations in the range $(0.2, 0.3)$. He warned that most generators of the form (8.1) will satisfy frequency and correlation tests over the complete cycle, but that a user should subject the output to a *runs test* (Exercise 8.3), since congruential generators tend to produce output with more runs longer than truly random choice would give. Empirical testing can eliminate poor choices of (a, b, m).

The Exercises illustrate other flaws in some approaches that are not immediately apparent. But the fact that a given congruential generator will always give the same sequence when using the same seed can be very useful—for games involving decisions during play, we can test out the consequences of different decisions for the same set of random events by running the game again from the same start point.

There are ways to modify the output of generators of the form (8.1) to make them more satisfactory. We could use several different generators, and cycle between them; or we might calculate blocks of 32 consecutive values, and use a different generator to 'randomly' shuffle them before use.

8.3 Codes and Ciphers

Why might *you* wish to encrypt a message? I assume you are not a diplomat sending sensitive information to your Government, or a military leader discussing your battle plan: such messages, if intercepted, could hail disaster. But when you buy goods or services over the internet, it is important that details of your debit or credit card are not read by a thief, who might use that information to use your card for his purchases. And if your email message is an enquiry about moving to another job, you might wish to conceal this fact from your current employer. *Cryptography* is the study of mathematical techniques in the field of information security.

Singh (1999) drew a distinction between codes and ciphers. In a cipher, individual letters or numbers are disguised, in a code whole words or phrases can be replaced by a single entity. Thus 'Operation Overlord' was the codeword used to define the Allied invasion of Normandy in 1944. If communications are to be in code, each party must possess a *code book* that lists all the code words that correspond to words we might use—loss or capture of this book would be a catastrophe. On the other hand, it is less vulnerable to frequency analysis of individual letters (see below). We look first at ciphers.

We refer to the actual message you wish to send, in words or numbers, as *plaintext*, and the version that is disguised for transmission as the *ciphertext*. Julius Caesar was an early exponent: in a *Caesar Cipher*, each letter in a word is replaced by another one which is, say, five steps further along the alphabet. So 'France' would become 'Kwfshj'; there are only 25 such ciphers, so trial and error will quickly yield the original message. Much better would be to use some random permutation of the letters—say A becomes U, B becomes L, C becomes T and so on. There are about 1.48×10^{26} such permutations in which no letter is coded to itself (Exercise 8.7) so, at the rate of one trial per second, it would take about 4.7×10^{18} years to run through all the possibilities. However, if the message is sufficiently long, knowledge of the different frequencies with which each letter appear can help decipher it relatively easily—in English, E turns up much more often that the other letters, followed by T, A, O, I, N and S. A computer can easily be programmed to make the respective counts. Another weakness is that the key, listing the actual permutation in use, might be intercepted. Some more secure method is required.

For safety, let pessimism rule. Bob and Alice should assume that their confidential messages to each other will be read by eavesdropper Eve. They seek *data security*—the messages cannot be changed during transit; *data authentification*—both can be confident that it really is the other person who sends the message; and *non-repudiation*—the sender cannot successfully deny that they were indeed the originator of the message. So they replace plaintext by ciphertext, and take steps to thwart Eve.

The original message may have both words and numbers, but we can always convert it to numbers only. The American Standard Code for Information Interchange (ASCII) assigns a 7-digit binary number to each letter. There are 128 such numbers, giving plenty of scope to encrypt punctuation marks and other useful symbols such a @. Converted to ASCII, a plaintext message will be a string of zeroes and ones, which can be broken down into blocks of convenient size, each block representing some number N. The successive values of N will be encrypted before transmission.

The central idea is that of a *key K*. With a Caesar Cipher, K states how far along the alphabet each letter is to be shifted. Bob and Alice must agree that key: for better security, even with a Caesar Cipher, they might move the first letter 5 steps along, the next 14 steps, the third 8 steps, and then repeat this 5–14–8 cycle over the whole message. This would give $25^3 = 15{,}625$ possible choices in which no letter is coded to itself, and $K = (5, 14, 8)$, which Eve must not know, or be able easily to discover. The larger the number of possible keys, the harder we expect it to be for Eve to succeed.

How can Bob and Alice agree on a key and keep Eve in ignorance? They might be able to meet each other, but perhaps Alice is in London and Bob in Cape Town; a trusted courier would be expensive, especially if they use a different key every time. So they seek a way to exchange messages in order to agree a key, but require that, even if Eve reads these messages, she cannot discover the key. Whitfield Diffie and Martin Hellman found a way to achieve this.

Alice/Bob openly state that the key will be the value of R^x (mod p), where R and the prime number p are public knowledge, but x will be known only to Bob and Alice. To establish x, Alice selects some value a with $1 \leq a \leq p - 1$, while Bob chooses b in the same range. They are kept secret, but neither a nor b should have a factor in common with $(p - 1)$—choosing them at random, with the techniques of the previous section, adds to the security. Easy-to-guess values such as birth dates should be avoided. Alice computes the value R^a (mod p) as α, Bob computes $\beta = R^b$ (mod p). These numbers α and β are exchanged, and it will not matter if Eve discovers them! For Alice can now compute β^a (mod p), Bob finds α^b (mod p), and these are equal since, modulo p,

Table 8.2 Computations of powers in modular arithmetic

Value of x	16	17	18	19	20	21
Value of $y = 3^x \pmod{79}$	16	48	65	37	32	17

$$\beta^a = (R^b)^a = R^{b \times a} \quad \text{and} \quad \alpha^b = (R^a)^b) = R^{a \times b}.$$

This common value is the secret key x. Knowing neither a nor b, Eve cannot compute x in the same way.

Example 8.4 Take $R = 3$ and $p = 17$, and suppose the secret choices are $a = 7$ and $b = 9$. Then, modulo 17, $3^7 = 11 = \alpha$, while $3^9 = 14 = \beta$; then $\beta^a = 14^7 = 6$ and also $\alpha^b = 11^9 = 6$, of course. Their private key is 6.

The crucial point is that, in practice, a gigantic value of p is used, so there is a vast quantity of possible values of $y = R^x \pmod{p}$. Even if Eve intercepts the value of y, and knows both R and p, she will find it infeasible to deduce x, as Example 8.5 illustrates.

Example 8.5 Take $R = 3$ and $p = 79$. Computing the values of $R^x \pmod{p}$ for $16 \leq x \leq 21$, we display the answers in Table 8.2.

You can see that, even if x_1 and x_2 are adjacent, then, unlike with ordinary arithmetic with real numbers, the corresponding values y_1 and y_2 might easily be far apart. So even if Eve knows y, and makes an initial guess x_0 for which 3^{x_0} is close to y, that is no guarantee that the correct value of x will be close to x_0. In effect, Eve is reduced to a brute force attack—she must run through all possible values of x in a systematic fashion, and only one of them will lead to y. On average, it takes about $p/2$ attempts to strike it lucky, and when p is enormous, say of order 10^{90}, even if a computer attack makes a billion tries every second, it would take on average over 10^{74} years to crack the code.

For large primes p and $R > 1$, the function $f(x) = R^x \pmod{p}$ can be termed a *one-way function*: given x, it is easy to compute $y = f(x)$, but given y, it is computationally infeasible to deduce x in a reasonable time. Bob and Alice can establish a secure key.

The RSA method of encryption (Rivest et al. 1978) relies on a different one-way function, the practical difficulty of factoring a very large number, known to be the product of two primes. You can swiftly deduce that $809 \times 997 = 806{,}573$, but if you simply knew that 806,573 was indeed the product of two primes it would take you some time to deduce them. (See Exercise 8.8.) Using RSA, Alice tells the world how to encrypt messages to her—only she will be able to decrypt them.

She selects two very large prime numbers p and q that she keeps secret. For illustration only, take $p = 3$ and $q = 17$, so that their product is $N = 51$. She finds the product of $(p - 1)$ and $(q - 1)$; for ease of reading, write $(p - 1) \times (q - 1) = r$, so that in this case $r = 2 \times 16 = 32$. She locates some number c that has no factors in common with r—here, $c = 7$ would work. She publishes the values of N and c to form her *public key*, here $(51, 7)$.

Bob wishes to send her a confidential message using her public key. He converts it via ASCII into a string of numbers $m_1, m_2, \ldots, m_k$, each of some given length. For any member M in this string, he calculates $M^c \pmod{N}$, giving the answer Z, say, and transmits the string $z_1, z_2, \ldots, z_k$. In our example, if his plaintext message is the single number $M = 4$, he calculates $4^7 \pmod{51}$. You should verify that the value, that he sends to Alice, is $Z = 13$. If Eve intercepts this, she knows how it was produced, but to go any further she needs to find p and q.

To decrypt the message, Alice has a *private* key, d, that she finds from her knowledge of p and q via the formula $c \times d = 1 \pmod{r}$. Recall that, in the Introduction above, the term *inverses* $\pmod{r}$ is used to describe this relation between c and d. Here, we seek d such that $7 \times d = 1 \pmod{32}$, and, since $7 \times 23 = 161 = 1 \pmod{32}$, we see that her private key is $d = 23$.

To read Bob's message, she computes $Z^d \pmod{N}$; here it is $13^{23} \pmod{51}$ which, as we calculated above in Example 8.1, is 4. That is indeed the message that Bob sent.

The reason that this process works comes from the 17th/18th century Fermat–Euler Theorem, that says that if p and q are different prime numbers, $N = p \times q$, $r = (p - 1) \times (q - 1)$, and m is not divisible by either prime, then $m^r = 1 \pmod{N}$. A proof is given in Churchhouse (2002).

Accepting the Theorem's validity, note that, modulo N, $Z^d = (M^c)^d = M^{c \times d}$. But the relation $c \times d = 1 \pmod{r}$ means that

$$c \times d = T \times r + 1$$

for some integer T. Hence

$$M^{c \times d} = M^{(T \times r + 1)} = (M^r)^T \times M.$$

But, mod N, the Fermat–Euler theorem shows that $M^r = 1$, from which $M^{c \times d} = (1^T) \times M = M$ follows.

We have glossed over one point—exactly how might Alice compute d from $c \times d = 1 \pmod{r}$ when r is enormous? An even older piece of Number Theory, the Euclidean Algorithm, paves the way. This is a systematic way to find the greatest common divisor (gcd) of two positive integers a and b with $a < b$. Plainly we can find Q (think *Quotient*) such that $b = Q \times a + R$ with $0 \le R < a$ (think *Remainder*). It follows that the gcd of a and b is the same as the gcd of R and a, so replace the pair (a, b) with (R, a), and continue this process until the result emerges. Computers working in integer arithmetic carry out this process swiftly and correctly, no matter how large the values of a and b. We illustrate with smaller numbers.

Example 8.6 To find the gcd of $a = 582$ and $b = 3018$, note that $3018 = 5 \times 582 + 108$, so that $Q = 5$ and $R = 108$. Then $582 = 5 \times 108 + 42$, after which we have

$108 = 2 \times 42 + 24$. Now $42 = 1 \times 24 + 18, 24 = 1 \times 18 + 6$, and finally $18 = 3 \times 6$. The gcd of 582 and 3018 is 6.

Example 8.7 Find the gcd of $a = 7$ and $b = 32$ via the Euclidean Algorithm, and *reverse* your working to find the inverse of 7 (mod 32).

Solution. Plainly $32 = 4 \times 7 + 4$, and then $7 = 1 \times 4 + 3$. Next $4 = 1 \times 3 + 1$, showing that $\gcd(7, 32) = 1$.

In reverse, $1 = 4 - (1 \times 3) = 4 - 1 \times (7 - 1 \times 4) = 2 \times 4 - 1 \times 7$. This last expression is also $(2 \times (32 - 4 \times 7) - 1 \times 7 = 2 \times 32 - 9 \times 7$. Rewrite this as $(-9) \times 7 = (-2) \times 32 + 1$, showing that the inverse of 7, (mod 32), is -9. However, modulo 32, we take $-9 = 23$, which falls in the range zero to 31.

Thus when $\gcd(c, r) = 1$, the reverse process shown in Example 8.7 illustrates how to find the inverse of c, mod r.

Using either Diffie–Hellman, or RSA, Bob and Alice can communicate with reasonable confidence that their messages are confidential. But what about the other desirable properties? How might we check that the message M has not been altered during its journey? One tool is to use a so-called *hash function*, often denoted by h. This will be a one-way function such as $h(M) = g^M$ (mod p) where p is a large prime—g can be small. The hash value of M is stored securely; at a later stage, the hash value of whatever M is believed to be can be found, and compared with this stored value, to check that M has not been changed.

To be sure the sender really is Alice, a MAC (Message Authentification Code) can also be transmitted. This is some number MAC(K, M) computed from the message M and an agreed key K. After composing M, Alice calculates this number, and sends it to Bob. On receipt, Bob decrypts to obtain M, then calculates the value of MAC(K, M) to see that it is the same as Alice sent.

In Sect. 4.5, we saw how check digits could help ensure that a codeword for a purchase or money transfer had been keyed in without error—*error detection*. Hill (1986) in his very readable account of the field of coding theory notes the use in Norway of two check digits for extra protection. Each citizen has an 11-digit registration code $x_1 x_2 \dots x_{11}$, where the first six digits give the date of birth (use DDMMYY) and the next three are allocated individually. Then x_{10} and x_{11} are chosen so that

$$3x_1 + 7x_2 + 6x_3 + x_4 + 8x_5 + 9x_6 + 4x_7 + 5x_8 + 2x_9 + x_{10} = 0 \quad (\text{mod } 11)$$

and

$$5x_1 + 4x_2 + 3x_3 + 2x_4 + 7x_5 + 6x_6 + 5x_7 + 4x_8 + 3x_9 + 2x_{10} + x_{11} = 0 \quad (\text{mod } 11)$$

both hold. Hill shows that the only double errors that go undetected are those where the digit in position 4 is incorrect by some quantity y 'too many', and that in position 10 is also incorrect, by this same quantity y 'too few'.

If, as well as detecting an error, we hope to correct it, a longer codeword is needed.

Example 8.8 We might use two binary digits to encode the four compass directions: 00 means North, 11 is South, 10 East and 01 West. A mistyping gives a valid, but incorrect code word. But extending each to three digits allows us to see that an error has been made—add an extra digit at the end of each codeword to give, respectively, 000, 110, 101 and 011. You should check that a single mistyping always gives an invalid code word.

Carefully add two more digits to each, giving 00000, 01101, 10110 and 11011. Now any single slip that gives a five-digit binary string leads to an invalid code word, but it differs from one of the valid code words in just one position, and in all the other valid codewords in at least two positions. So we can correctly interpret it as being the closest valid codeword, and 'correct', as well as detect, the error.

8.4   Search Engines

We are now used to the idea that we can find information about some topic—the Crimean War, vegan diets, the current weather in Rome—by simply typing a relevant word or phrase into a computer search engine. Almost instantly, a list of pertinent web pages appears on the screen. How are they chosen, and why are they in a particular order?

The best-known search engine, the *PageRank Algorithm* was developed by Larry Page and Sergey Brin. Suppose there are N web pages that are potentially relevant to our topic. Typically N is several million, but to see what is happening, consider an example where $N = 4$. To rank these pages in order of their *importance*, count how often each page refers to another relevant page. For example, suppose page 1 refers to all the other pages, page 2 refers just to pages 3 and 4, page 3 refers only to page 1, while page 4 refers to just pages 1 and 3. The more often a page is referenced, the more important it is assumed to be.

Let $\mathbf{u} = (u_1, u_2, u_3, u_4)^T$ be some non-negative column vector, whose components are thought to measure the relative importance of each page. The links are assumed to transfer a page's importance to those pages it refers to in equal amounts: so page 1 gives $1/3$ of its importance to each of pages 2, 3 and 4; pages 2 and 4 transfer half their importance to two other specified pages; page 3 transfers all its importance to page 1. This is summarised in the 4×4 matrix

$$A = \begin{pmatrix} 0 & 0 & 1 & 1/2 \\ 1/3 & 0 & 0 & 0 \\ 1/3 & 1/2 & 0 & 1/2 \\ 1/3 & 1/2 & 0 & 0 \end{pmatrix}.$$

If $\mathbf{u}$ does genuinely measure the importance of each page then, after the transfers have been made, we hope to end up with $\mathbf{u}$ again. In symbols, we hope to have $A\mathbf{u} = \mathbf{u}$. Any multiple of a solution is also a solution, so for uniqueness we could ask that the sum of the components of **u** is unity. In our example, you can verify that **u** = $(12, 4, 9, 6)/31$ is the solution, leading to the rank order 1, 3, 4, 2.

In this instance, up to the scaling factor there is a unique solution, but this does not always occur. There might be eight relevant pages, but no references at all between the first five and the last three. Then A would have the format

$$A = \begin{pmatrix} B & C \\ D & E \end{pmatrix}$$

where B is 5×5, E is 3×3, while C is a 5×3 matrix of zeroes, and D is a 3×5 matrix of zeroes. If we can find non-negative $\mathbf{u} = (u_1, u_2, u_3, u_4, u_5)^T$ with $B\mathbf{u} = \mathbf{u}$, and $\mathbf{v} = (v_1, v_2, v_3)^T$ with $E\mathbf{v} = \mathbf{v}$ then, with t such that $0 < t < 1$, write $\mathbf{w} = (tu_1, tu_2, tu_3, tu_4, tu_5, (1-t)v_1, (1-t)v_2, (1-t)v_3)^T$. Clearly, $\mathbf{w}$ is non-negative and $A\mathbf{w} = \mathbf{w}$. We can rank the first five and the last three pages among themselves, but there is no unique way to intertwine them to rank all eight pages.

Another issue is that there may be a relevant page that has no links to any other page—there is nowhere to transfer its importance to. To overcome these problems, let B be the $N \times N$ matrix all of whose entries are $1/N$, and let p be some constant with $0 < p < 1$. Now define M as the $N \times N$ matrix $M = (1 - p)A + pB$. (p is referred to as the *damping factor*, typically taken to be 0.15.) Every entry of M is positive, the column sums of M are unity, and in these circumstances the celebrated Perron–Frobenius Theorem (met in Chap. 3) tells us there is a vector **u**, unique up to a scaling factor, with $M\mathbf{u} = \mathbf{u}$. This leads to a ranking of all the relevant pages, the absence of links has been overcome.

There is a useful interpretation of M and **u**, when the components of **u** are chosen to sum to unity. Imagine a fly hopping among the N relevant pages; from any page, with probability $(1 - p)$ it jumps to one of the pages it is linked to, selected at random; and with probability p, it jumps to some page chosen completely at random. In the long run, the fly will be at page j with probability u_j. The more important the page, the more often the fly visits it.

We should check how invoking M alters the original outcome for the example matrix A above. For arithmetical convenience, take $p = 0.16$, so that

$$M = \begin{pmatrix} 0.04 & 0.04 & 0.88 & 0.46 \\ 0.32 & 0.04 & 0.04 & 0.04 \\ 0.32 & 0.46 & 0.04 & 0.46 \\ 0.32 & 0.46 & 0.04 & 0.04 \end{pmatrix}.$$

Rounding the exact solution of $M\mathbf{u} = \mathbf{u}$ to two significant figures we obtain $(0.37, 0.14, 0.29, 0.20)$, compared to $(0.39, 0.13, 0.29, 0.19)$ from $A\mathbf{u} = \mathbf{u}$. The rank order is identical.

With millions of pages to rank, solving the vast system of simultaneous equations directly would be infeasible. But an iterative scheme leads quickly to a ranking. Begin with an initial guess that all pages are equally important, represented by the column vector **e** all of whose entries are $1/N$, and update it by computing $M\mathbf{e}$. The components representing pages with many incoming links (proxies for its importance) will increase: continue by finding $M(Me) = M^2e$, $M(M^2)e = M^3e$ and so on—fast computations. You should check that, after just one iteration in our example to find Me, the correct rank order is already established. This method relies on the Perron–Frobenius Theorem, that assures us that if M is a $N \times N$ matrix with all entries positive and column sums unity, and e is as described, then the sequence of vectors $(M^k e : k = 1, 2, 3, \ldots)$ converges to the unique vector u which satisfies $Mu = u$, whose entries sum to unity.

So given A and a choice of damping factor, we find M and perform this sequence of iterations. This will establish the order of importance. Even better, as pages build up a history which picks out those often considered important, we can begin the iteration with a vector u expected to be closer to the limiting value.

8.5   Exercises

8.1 John von Neumann (1951) suggested the *mid-square* method of producing random digits: take an initial 4-digit number, e.g. 5379 and square it to get 28,933,641; select the middle four digits, 9336 to begin the sequence as 9, 3, 3, 6, and square 9336 to get 87,160,896, leading to 1, 6, 0, 8 and so on. (Since 1608^2 has only seven digits, insert a leading zero.) Plainly this must repeat after at most 10,000 iterations: illustrate its deficiencies by using the initial values 8414 and 9999.

8.2 For the congruential generator $x_{n+1} = 3x_n + 4$ (mod 17), find the cycle length for each possible seed s with $0 \leq s \leq 16$.

8.3 Given any sequence of different real numbers, we define a *run up* as a subsequence of consecutive numbers that fall in increasing order. So any such sequence can be split into a succession of runs up: for example, with 15904237895214689 we can insert dividing lines to produce 159|04|23789|5|2|14689| which has *runs up* of respective lengths 3, 2, 5, 1, 1, 5. Runs of any length are plainly possible. Suppose we have a genuinely random $U(0, 1)$ sequence of values.

(a) Show that the probability that the length of the first run up is at least r is $1/r!$.

(b) Deduce that the probability the length is exactly r is $1/r! - 1/(r + 1)!$.

(c) Informally, explain why the lengths of consecutive runs up are *not* independent but that, if we discard the first value after each run up and then consider consecutive lengths, they *are* independent.

(d) Hence explain how a satisfactory *runs test* for a linear congruential generator might be constructed. (There is no need to go into statistical detail.)

8.4 A Fibonacci-type recursion takes two initial integer values x_0 and x_1 and, for $n \geq 0$, uses $x_{n+2} = (x_{n+1} + x_n)$ (mod m) for some suitable integer m to generate a sequence. Write $v_n = x_n/m$ as a putative pseudorandom $U(0, 1)$

sequence. Show that it is impossible that $x_n < x_{n+2} < x_{n+1}$, and say why this renders the method unsuitable for its alleged purpose.

8.5 Suppose $(U_n : n = 1, 2, 3, \ldots)$ and $(V_n : n = 1, 2, 3, \ldots)$ are independent random sequences of $U(0, 1)$ values. Define the sequence $(W(n))$ by taking $W_n = U_n/2$ if n is odd, and $W_n = (1 + V_n)/2$ if n is even. Show that, if n is selected to be odd or even at random, then W_n is $U(0, 1)$, but that the sequence $(W_n : n = 1, 2, 3, \ldots)$ is not a $U(0, 1)$ sequence.

8.6 You intercept the message QIIX QIEX QMHR MKLX which you (correctly) take to have been produced as a Caesar Cipher. Decrypt it, and, using the same cipher, reply 'agreed'.

8.7 Let $D(n)$ be the number of permutations of the integers $(1, 2, \ldots, n)$ in which no member is in its correct position (a *derangement*). Show that $D(1) = 0$, $D(2) = 1$ and, when $n \geq 2$, $D(n + 1) = n(D(n) + D(n - 1))$. Use induction to show that $D(n) = n!(1 - 1 + \frac{1}{2!} - \frac{1}{3!} + \cdots + (-1)^n \frac{1}{n!})$ for $n \geq 1$.
Hence verify the claim that there are about 1.48×10^{26} derangements of an alphabet of 26 letters.

8.8 Take up the challenge in the text—given that 8,023,380,749 is the product of two primes, find their values.

8.9 Alice and Bob will use the Diffie–Hellman method to establish a key K using $3^x \pmod{17}$.

(a) To agree to the value of K, Alice generates her private key at random, selecting 7. What is her public key?

(b) Bob's public key is 4. What is his private key?

(c) Give their respective computations, and show that both lead to the same value for K. State this value.

8.10 Show that gcd(259, 5412)=1. Working backwards, find the inverse of 259, modulo 5412.

8.11 For the RSA method, suppose Alice selects $N = 77$ and $c = 43$ as her public key.

(a) Factorise 77 as $p \times q$, and show that c has no factor in common with $(p - 1) \times (q - 1)$.

(b) Deduce the value of her private key.

(c) Bob sends her a message M, which arrives encrypted to $C = 5$. Deduce M.

8.12 A search engine uses the PageRank algorithm, without damping, on a topic with just four web pages that reference it. Page 1 refers only to page 2, page 2 refers to both pages 1 and 3, page 3 links to pages 2 and 4, while page 4 refers only to page 1. Estimate the relative importance of each page, and rank the pages in presumed order of importance.

References and Further Reading

Churchhouse R F (2002) Codes and Ciphers. Cambridge University Press
Diffie W and Hellman M E (1976) New Directions in Cryptography. IEEE Transactions on Infor-
    mation Theory 22(6) 644–54
Hill R (1986) A First Course in Coding Theory. Oxford University Press
Knuth D E (1981) The Art of Computer Programming, Vol 2 (2nd Edition) Addison-Wesley
Morgan B J T (1984) Elements of Simulation. Chapman and Hall
Rivest R L, Shamir A and Adelman L (1978) A method for obtaining digital signatures and public
    key cryptosystems. Communications of the Association for Computing Machinery Vol 21(2)
    120–126
Singh S (1999) The Code Book. Fourth Estate
von Neumann J (1951) Various techniques used in connection with random digits. Monte Carlo
    Method, National Bureau of Standards Applied Mathematics Series 12 36–38

Appendix
Useful Mathematical Facts

1. $1 + 2 + 3 + \cdots + n = n(n+1)/2$.
2. $1^2 + 2^2 + 3^2 + \cdots + n^2 = n(n+1)(2n+1)/6$.
3. $\binom{n}{r} = \frac{n!}{r!(n-r)!}$ is the number of ways of choosing r objects from a collection of n objects, without regard to their order.
4. $(a+b)^n = \sum_{r=0}^{n} \binom{n}{r} a^{n-r} b^r = a^n + na^{n-1}b + \binom{n}{2}a^{n-2}b^2 + \cdots + b^n$.
5. For $|x| < 1$, then $\log(1+x) = x - \frac{x^2}{2} + \frac{x^3}{3} - \frac{x^4}{4} + \cdots$.
6. (a) Sum of a finite geometric series, $\sum_{k=0}^{n-1} x^k = \frac{1-x^n}{1-x} = \frac{x^n-1}{x-1}$ if $x \neq 1$.

   (b) When $|x| < 1$, since $x^n \to 0$ as $n \to \infty$, we obtain

   (i) $\sum_{k=0}^{\infty} x^k = 1/(1-x)$.　　　　Then, by differentiation,

   (ii) $\sum_{k=1}^{\infty} kx^{k-1} = 1/(1-x)^2$.　　Multiply by x to see that

   (iii) $\sum_{k=1}^{\infty} kx^k = x/(1-x)^2$.

7. For fixed x, $\lim_{n \to \infty}(1 + \frac{x}{n})^n = e^x = \exp(x)$.
8. For all x, $\exp(x) = e^x = 1 + x + \frac{x^2}{2!} + \frac{x^3}{3!} + \cdots$.
9. To solve $f(x) = 0$ via a *Newton–Raphson* iteration, make an initial guess $x_0 = c$, then calculate $x_1, x_2, \ldots$ from $x_{n+1} = x_n - f(x_n)/f'(x_n)$. If the sequence converges to a limit, that limit is a solution.
10. To solve $f(x) = 0$ using the *Midpoint Rule*, suppose x_0 and x_1 are such that $f(x_0) < 0$ and $f(x_1) > 0$. Write $x_2 = (x_0 + x_1)/2$ (the midpoint of x_0 and x_1), and evaluate $f(x_2)$. If $f(x_2) > 0$, take $x_3 = (x_0 + x_2)/2$, while if $f(x_2) < 0$, take $x_3 = (x_1 + x_2)/2$, and so on. Provided f is a continuous function, the sequence $\{x_i\}$ converges to a solution. Proceed in an analogous fashion if $f(x_0) > 0$ but $f(x_1) < 0$.

© Springer Nature Switzerland AG 2019
J. Haigh, *Mathematics in Everyday Life*,
https://doi.org/10.1007/978-3-030-33087-3

11. To solve $x = g(x)$ via an *iteration scheme*, let x_0 be an initial guess, and take $x_{n+1} = g(x_n)$ for $n \geq 0$. Provided that x_0 is close enough to the answer, *and* $|g'(x)| < 1$ for x near the desired root, the sequence will converge as required. (We could convert the equation $f(x) = 0$ to this format by *defining* $g(x) = f(x) + x$.)

12. *L'Hôpital's Rule* deals with expressions of the form $f(x)/g(x)$ at any point $x = c$ where $f(c) = g(c) = 0$. Assuming all expressions make sense, the Rule states that, if $f(c) = g(c) = 0$, then, as $x \to c$, so $f(x)/g(x) \to f'(c)/g'(c)$. If you find that we also have $f'(c) = g'(c) = 0$, worry not: apply the Rule to $f'(x)/g'(x)$, so that the desired limit is just $f''(c)/g''(c)$—etc.

13. Suppose that f is a function that can be differentiated as often as we like. Then its *Maclaurin expansion* is

$$f(0) + xf'(0) + \frac{x^2}{2!} f''(0) + \frac{x^3}{3!} f'''(0) + \cdots .$$

Its *Taylor expansion* about the point c is

$$f(c) + (x - c)f'(c) + \frac{(x - c)^2}{2!} f''(c) + \frac{(x - c)^3}{3!} f'''(c) + \cdots .$$

Provided these series converge, the first few terms can be expected to approximate f well for values near zero, or near c, respectively.

14. Here, informally, is how the Maclaurin expansion arises. (A similar argument works for the Taylor series.) SUPPOSE there are coefficients $a_0, a_1, a_2, \ldots$ such that

$$f(x) = a_0 + a_1 x + a_2 x^2 + a_3 x^3 + \cdots ,$$

and that everything behaves nicely. Put $x = 0$; then $f(0) = a_0$, as all other terms are eliminated. So IF such an expansion exists, then $a_0 = f(0)$. Differentiate to see that $f'(x) = a_1 + 2a_2 x + 3a_3 x^2 + \cdots$; put $x = 0$ again, now obtaining $a_1 = f'(0)$. Differentiate again, find $f''(x) = 2a_2 + 6a_3 x + \cdots$, and again put $x = 0$. This gives $f''(0) = 2a_2$, i.e. $a_2 = f''(0)/2$. Keep on taking the tablets. This is NOT a proof that a function is equal to its Maclaurin expansion. The logic merely says 'IF there is a power series expansion, THEN it has this Maclaurin form'. But there may be no such expansion: for example, when $f(x) = |x|$.

15. We can use Taylor expansions to see how L'Hôpital's Rule arises. Assume that the functions $f(x)$ and $g(x)$ have valid Taylor series about the point c, and that $f(c) = g(c) = 0$. Then we have

$$\frac{f(x)}{g(x)} = \frac{f(c) + (x - c)f'(c) + \frac{(x-c)^2}{2!} f''(c) + \cdots}{g(c) + (x - c)g'(c) + \frac{(x-c)^2}{2!} g''(c) + \cdots}$$

But $f(c) = g(c) = 0$, and when $x \neq c$, we can cancel the common factor $(x - c)$ in every other term. Thus numerator and denominator simplify leading to

$$\frac{f(x)}{g(x)} = \frac{f'(c) + \frac{(x-c)}{2}f''(c) + \frac{(x-c)^2}{6}f'''(c) + \cdots}{g'(c) + \frac{(x-c)}{2}g''(c) + \frac{(x-c)^2}{6}g'''(c) + \cdots}.$$

Provided not both of $f'(c)$ and $g'(c)$ are zero, then, as $x \to c$, so this ratio converges to $f'(c)/g'(c)$. And if $f'(c) = g'(c) = 0$, then $(x - c)/2$ is a common factor in all other terms; cancel throughout obtaining

$$\frac{f(x)}{g(x)} = \frac{f''(c) + \frac{x-c}{3}f'''(c) + \cdots}{g''(c) + \frac{x-c}{3}g'''(c) + \cdots}.$$

Again, if not both $f''(c)$ and $g''(c)$ are zero, we can make sense of this as $x \to c$, and so on.

(This is not, of course, an actual proof of L'Hôpital's Rule. Even if the Taylor series is valid, we have to take care in replacing $(x - c)$ by zero, *infinitely often as $x \to c$*.)

16. Useful approximations include: $2^{10} \approx 1000$; $\exp(3) = e^3 \approx 20$; $\pi^2 \approx 10$.
17. (a) $1 + \frac{1}{2} + \frac{1}{3} + \cdots + \frac{1}{n} \approx \log(n) + 0.577$.
    (b) $1 + \frac{1}{2^2} + \frac{1}{3^2} + \cdots = \frac{\pi^2}{6}$.
18. If you think you have run out of symbols, here is the Greek alphabet:
    $\alpha, \beta, \gamma, \delta, \epsilon, \zeta, \eta, \theta, \iota, \kappa, \lambda, \mu, \nu, \xi, o, \pi, \rho, \sigma, \tau, \upsilon, \phi, \chi, \psi, \omega.$

Index

Printed in the United States
By Bookmasters